ROUTLEDGE LIBRARY EDITIONS: ARISTOTLE

Volume 3

ARISTOTLE AND PHILOPONUS ON LIGHT

ARISTOTLE AND PHILOPONUS ON LIGHT

JEAN DE GROOT

Routledge
Taylor & Francis Group
LONDON AND NEW YORK

First published in 1991

This edition first published in 2016
by Routledge
2 Park Square, Milton Park, Abingdon, Oxon OX14 4RN

and by Routledge
711 Third Avenue, New York, NY 10017

Routledge is an imprint of the Taylor & Francis Group, an informa business

British Library Cataloguing in Publication Data
A catalogue record for this book is available from the British Library

ISBN: 978-1-138-92762-9 (Set)
ISBN: 978-1-315-67490-2 (Set) (ebk)
ISBN: 978-1-138-93705-5 (Volume 3) (hbk)
ISBN: 978-1-138-94235-6 (Volume 3) (pbk)
ISBN: 978-1-315-67492-6 (Volume 3) (ebk)

Publisher's Note
The publisher has gone to great lengths to ensure the quality of this reprint but points out that some imperfections in the original copies may be apparent.

Aristotle and Philoponus on Light

Jean De Groot

Garland Publishing, Inc.
New York & London
1991

Library of Congress Cataloging-in-Publication Data

De Groot, Jean.
Aristotle and Philoponus on light / Jean De Groot
p. cm. — (Harvard dissertations in the history of science)
Revision of the author's thesis (Ph. D.)—Harvard University, 1979.
Includes bibliographical references and index.
ISBN 0-8240-7251-0 (alk. paper)
1. Philoponus, John 6th cent.—Contributions in optics. 2. Aristotle—Contributions in optics. 3. Optics—History. I. Title. II.Series.
QC381.D39 1991
535—dc20 91-36012

Printed on acid-free, 250-year-life paper

Manufactured in the United States of America

To Simon

Contents

Preface

This book is substantially the same as my 1979 doctoral thesis completed for the History of Science Department at Harvard University. Philoponus' long commentary on Aristotle's definition of light (*De Anima* II.7, 418b9) sets up the major concerns, both in optics and theory of light, that the book treats. This long commentary is translated in Appendix I. The passage illustrates how light was of special interest in Neoplatonism because of its being something incorporeal in the world of natural bodies. Due to its incorporeality, light had a special role to play in the philosophical analysis of the interpenetration of bodies and also as a paradigm for the soul-body problem.

Besides containing much interesting material concerning the physiology of vision, this passage also contains an account of the propagation of light in no time whatsoever. The philosophical background of this theory is the subject of Chapters Three through Five. These chapters detail the philosophical origins of the causal structure common to impetus theory and to the theory of light that came to be called "multiplication of species." Investigating the background of these theories in Aristotle's philosophy shows that they are integrally involved in issues of perennial interest in interpreting Aristotle. Specifically, these issues in the history of science are related to Aristotle's distinction between *kinêsis* and *energeia*, and Chapters Three through Five can be regarded as an analysis of the development and application of that distinction in Alexander, Proclus, and Philoponus. Essential aspects of that analysis appeared in my 1983 article, "Philoponus on *De Anima* II.5, *Physics* III.3, and the Propagation of Light" (*Phronesis* 28, no.2 (1983).

Philoponus' commentary on Aristotle's definition of light also presents an interpretation of mathematical science which is of great interest.

Philoponus said it makes no difference from the point of view of geometrical optics whether the lines of the visual cone proceed from the object to the eye or *vice versa*. Chapters Six and Seven examine Philoponus' view of the physical basis for mathematical demonstration that supports this belief. Chapter Six incorporates material that did not appear in my doctoral thesis. This new material comes from the commentary on *Posterior Analytics* II attributed to Philoponus and concerns ideas about the relation between conceivability and existence in the analysis of definition for physical things. The conceivability of secondary substratum, Philoponus' three-dimensional corporeal extension, forms an important part of Philoponus' argument for its priority to substance.

Much interesting work on Philoponus and the other Aristotelian commentators has been done in the last ten years, much of it through the efforts of Richard Sorabji. His three books, *Philoponus and the Rejection of Aristotelian Science*, *Matter, Space, and Motion*, and *Aristotle Transformed*, are the three best reference books in Philoponus studies. *Aristotle Transformed* contains a contribution by Koenraad Verrycken, "The Development of Philoponus' Thought and its Chronology," which adds much to our still fragmentary understanding of Philoponus' intellectual history and his relation, as a Christian, to the school of pagan philosophy at Alexandria. My own Chapter One presents a summary of the purely historical information about the survival of the school of pagan philosophy at Alexandria, set in the context of the wider political problems of which the Alexandrian patriarchate formed the center. My own view is that the survival of the school in Alexandria is not of great importance for understanding the status of pagan philosophy at this date, but was a small facet of the generally unruly cultural situation in Alexandria. Verrycken's belief that there was no specific conversion date in Philoponus' philosophical career is a view I share.

A pleasure associated with publishing a doctoral thesis is having the opportunity to thank those whose generosity and interest helped launch a young scholar. It would be difficult to acknowledge the assistance of all the people who contributed to my thesis by their suggestions, criticism, and discussion. However, I would like to thank those whose help most directly influenced its development.

I am indebted to M. Pierre Hadot, Directeur d'études, École Pratique des Hautes Études, V^e^ section, for encouraging me at an early stage of my research and for introducing me to his younger colleague, M. Alain Segonds. M. Segonds gave most generously of his time and knowledge, as

I learned to read late antique Greek and understand its philosophical vocabulary. I am grateful to R.W. Sharples for reading Greek with me while I was in England, and to John Duffy for the opportunity of reading the entire *Timaeus* with him during my first year at Dumbarton Oaks. I also thank John for reading my translation of Philoponus' commentary on *De Anima* II.7, 418b9, and suggesting improvements. I was able to consult with all these people and to work full-time on my thesis, because of fellowship support from Harvard University and Dumbarton Oaks.

Closer to home, I wish to thank my advisors, Professors John E. Murdoch and A.I. Sabra, who offered both encouragement and constructive criticism throughout the period of my thesis research. Their comments enabled me to improve my thesis considerably during the year before its submission. I am especially grateful to Martha Nussbaum, who read both drafts of my thesis, pointing out specific problems and making suggestions for improvement. Finally, I am grateful to Eunice Rice for typing the manuscript onto disk for me so that this publication of it would be possible, and to Maridella Hegarty for technical assistance in the School of Philosophy at Catholic University.

I thank my husband, Simon De Groot, for proof-reading the final version and for all his encouragement in moving this project to fruition. This book is dedicated to him.

Aristotle and Philoponus on Light

Chapter One

Philoponus and Alexandria: An Historical Introduction

In 529, Justinian ordered that the school of pagan philosophy in Athens be closed. Teaching continued at a similar school in Alexandria until well into the seventh century. There has been considerable modern speculation about the significance of these events in Athens and Alexandria for evaluating the political and intellectual status of paganism at this late date in the Roman Empire. The situation is especially important for interpreting the role of philosophy in the conflict between paganism and Christianity, and for evaluating the extent to which pagan philosophy had been assimilated by Christian thinkers. For understanding the relation of Christianity to pagan philosophy, the role of John Philoponus in the school at Alexandria is of considerable importance. Philoponus, though a Christian, edited many of the commentaries of Ammonius, head of the Alexandrian school. In some cases, he made changes and added comments of his own. There were many Christian students in Alexandria who attended the classes of Ammonius, before going on to study law or rhetoric. However, Philoponus is the only Christian of whom we have knowledge, either at Athens or Alexandria, who pursued the masters of pagan philosophy so thoroughly as to publish commentaries under his own name. Consequently, his status in the school has puzzled scholars who have attempted to explain the survival of pagan philosophy at Alexandria.

This issue is of interest to the historian of science or philosophy because of what it might reveal about Philoponus' attitude toward the pagan Neoplatonism that dominated late antique school philosophy. The historical and intellectual aspects of Philoponus' relationship to the pagan philosophical tradition are closely allied, because of the nature of the commentary genre. The commentaries are the artifacts of the school tradition. They are our primary source of historical information about the schools, but at the same time, they embody the intellectual tradition, which we wish to evaluate for its own sake. In Philoponus' case, separating the substance of original thought from the conventions of the lecture form is complicated by the difficulty of determining whether the ideas presented belong to him or to his teacher, Ammonius. Consequently, the historical interpretation of Philoponus' relation to the Alexandrian school impinges upon the interpretation of his philosophical works. For this reason, the survival of the school of Alexandria, after the school at Athens had been closed, is an event of considerable interest to any student of late antique philosophy. A brief evaluation of the circumstances surrounding the survival of the school is therefore in order.

According to the testimony of the sixth century chronicler, John Malalas, in 529 Justinian sent an edict (*prostaxis*) to Athens forbidding the teaching of philosophy or law there.[1] There are a number of questions about Malalas' report. In the first place, beside the evidence of this report, there is no reason to believe there was a law school at Athens. Secondly, there is nothing in Malalas' report that suggests the edict was aimed at the pagan school of philosophy or at pagans in general. Remarks of members of the Academy at this time indicate they were refused the right to teach due to their pagan convictions.[2] However, if we assume the order was directed at the Neoplatonic Academy, it is difficult to understand why a particular edict was needed for them, since general laws against pagans were quite sufficient to outlaw their activities.[3] It is possible that these general laws were more or less ineffective at a local level without specific imperial attention, and this may explain the special edict for Athens. However, in Athens, even after 529, Simplicius continued his philosophical research, and it is likely that he used the resources of the Academy library. Furthermore, Olympiodorus, writing after 529, said that despite many confiscations, enough of the school's endowment remained to pay the *diadochos*.[4] All we can say with assurance is that teaching ceased at the Academy in 529, ostensibly as a result of the special edict Malalas mentioned.

Whatever the nature of the special edict against Athens, it seems clear that the general laws against pagans should have applied as well to Alexandria as Athens. Ammonius, who held the chair of philosophy in Alexandria during this period, was pagan, as were his successors, Eutocius and Olympiodorus. The first Christian head of the school was Elias,

Olympiodorus' successor. Elias became *philosophos* at least forty years after teaching stopped at the Athenian school. Pagans had, therefore, been able to continue teaching philosophy in Alexandria in spite of the laws against them. And unlike Athens, once the pagan succession stopped, teaching did not cease, but rather was carried on by Christians.[5]

The Athenian philosophers testify that their academic activities were restricted because of their pagan beliefs. Athenian Neoplatonism was associated by its practitioners, as well as its opponents, with pagan religion.[6] This association is lacking in the Alexandrian school of philosophy. We cannot be sure that the Alexandrians abjured the Iamblican Neoplatonism of Athens, with its theurgic practices, since we do not have their writings on metaphysics. However, we do have the testimony of Damascius, last *diadochos* at Athens, that Aristotle was the special interest of Ammonius, his counterpart in Alexandria.[7] The professional philosophy of Aristotle lacked the religious overtones of Iamblichan Neoplatonism. Even Christians recognized the importance of Aristotelian logic, although they were uncomfortable with his natural philosophy.[8] Damascius accused Ammonius of having compromised with the Christian authorities, and modern scholars have supposed this to be a reference to the conditions under which the Alexandrian school was allowed to continue.[9] Damascius may have assumed the consummation of a deal where there existed, in fact, only an informal arrangement, based on a long-standing agreement. Nevertheless, there has been considerable speculation about the nature of a possible compromise between Ammonius and the Christian authorities.

The necessity for student contributions has been presented by both Tannery and Saffrey as partial reason for a compromise that Ammonius supposedly made with the patriarch, Athanasius. According to Tannery, for Ammonius to obtain Christian students

> ...il fallu que les maîtes se pliassent à certaines nécessités; ils durent éviter de blesser les croyances de leurs élèves et tout aussi bien se garder que l'évêque n'interdît leurs leçons à ses fideles.[10]

Saffrey endorsed this view and further suggested that Ammonius may have agreed to limit his exegesis to Aristotle, since Plato was probably less acceptable to the Christian authorities.[11] No doubt, there were Christian students in Ammonius' classes. Zacharias of Mitylene, in his dialogue, *Ammonius*, recounted an exchange between Ammonius and some of his Christian students on the eternity of the world during the course of a lecture on the *Physics*. From Zacharias' description of his own activities, Courcelle supposed that Zacharias meant the episode to be taking place in 486.[12] However, Saffrey does not even suggest that Ammonius made his deal with the patriarch until 495. Zacharias' dialogue tells us that Christians were present in Ammonius' classes long before the supposed compromise. Consequently, for this hypothesis to be plausible, Athanasius must have threatened to *remove* the paying Christian students from Ammonius' lectures, or to use his influence with the imperial authorities in the city to prevent Ammonius from teaching. The latter alternative is the more reasonable choice. In any case, Athanasius' tenure as patriarch ended long before Justinian's laws against pagans took effect. Even if Ammonius made a compromise of sorts, it was not a response to the pagan persecutions that closed the Academy. Rather, such a compromise would reflect circumstances peculiar to Alexandria.

Damascius' charge that Ammonius compromised with the authorities at least strengthens the very plausible idea that Ammonius' attitude toward his Christian antagonists was relevant to the survival of his school. Some idea of Ammonius' attitude toward Christian doctrines can be gained from Zacharias' dialogue. He presented several discussions he had with Ammonius in class, concerning doctrines of Christianity such as the creation of the world and the Trinity.[13] Ammonius appeared interested in Christian philosophical arguments and, to a certain extent, impressed by their cogency. From the dialogue, it appears that, unlike his Athenian counterparts, Ammonius was no antagonist to Christian doctrine. His own commitment to pagan philosophy did not make him intolerant of Christian arguments on the same issues, and he was willing to consider those arguments on their merits. Whether or not this tolerance was the result of necessity, it is nevertheless true that in practical terms Ammonius lacked the commitment to pagan culture as such which was always the natural concomitant of the teaching of philosophy in Athens. This conciliatory attitude is undoubtedly the heart of the *homologia* with which Damascius reproached his former teacher.

Although Ammonius' attitude toward Christianity is undoubtedly important, it is also worthwhile to note the larger political climate of Alexandria that may have affected the survival of his school. Ever since Justin had become emperor in 518, it had been imperial policy to enforce in the east the results of the Council of Chalcedon (451) and to end the schism between the eastern and western churches. The Council of Chalcedon declared that Christ was of two natures, human and divine. However,

Christians in Egypt and Syria had long held to the view that Christ had a single nature, the divine, and they were not prepared to renounce their view in favor of another, which they considered non-orthodox. Compromising additions to the two-nature statement, which might have brought the eastern churches into the fold, were not adopted by the Council. Both of the eastern emperors, Zeno and Anastasius, were sympathetic to the Monophysite cause. It was not until Justin became emperor that a policy of enforcing orthodoxy was established. Monophysite patriarchs held the Sees of Antioch and Alexandria. Severus, patriarch of Antioch, was replaced by an orthodox patriarch in 518. However, a Monophysite, Timothy IV, continued as patriarch in Alexandria, having succeeded Dioscorus II in 517. In fact, it was to Alexandria that Severus fled in 518, and there he remained the most powerful leader in the Monophysite movement.

During the reign of Justin, Egypt was excluded from the campaign of persecution against Monophysites throughout the rest of the eastern provinces.[14] This imperial exemption was, for Alexandria, exercised in other matters as well. In 520-21, Justin put an end to the Olympic games in cities throughout the Empire and also to other spectacles and entertainments because of the riots caused by political factions. Such entertainments were allowed to continue in Alexandria, and A.A. Vasiliev suggested why:

> The Alexandrian spectacles remained undisturbed by imperial decrees and continued to amuse and divert the population of the Egyptian capital, whose tranquillity and satisfaction were essentially important for the tranquillity and prosperity of the empire as a whole.[15]

Contributing to the economic reasons for maintaining tranquillity in Egypt was the deep-rooted Monophysism of Egyptian Christians, especially the monks, and the political power that traditionally resided in the Alexandrian patriarchate.[16] Furthermore, by this time, the Alexandrian patriarchate was the rallying point for the Monophysite population still clinging to what they regarded as the orthodox faith. Both Justin and Justinian wished for a final reconciliation of the Monophysites to orthodoxy, but they both perceived the peril of insurrection that lay in attempting to force that reconciliation. Concessions were made to maintain peace in Egypt. Not all were strictly related to religious matters; some were aimed at placating the general population. The proscription against teaching by pagans, therefore, was not the only imperial edict not enforced in Alexandria, and it certainly was not the most important law that was ignored. The inability of the emperor to enforce his will in Egypt may have resulted in the survival of the school, either by negligence or because the Monophysites wished to see the school continue. At this juncture, the presence of Christian students in Ammonius' classes may well have been important. The association of Philoponus with the school suggests its value to the Monophysites, beyond its function in training students destined for law or rhetoric. Philoponus was a Monophysite with a great interest in the application of Aristotelian philosophy to theology. The Monophysites needed to substantiate their position from a theological as well as a political standpoint.

When Saffrey presented his hypothesis of the agreement made between Ammonius and the Christian authorities, he also suggested that Philoponus was Ammonius' appointed assistant. He supposed that there was greater safety for Ammonius in publishing his lectures under the name of a Christian. Furthermore, he thought it was no coincidence that Philoponus' treatise, *De Aeternitate Mundi Contra Proclum*, was published in the same year that teaching ceased at the Neoplatonic Academy. He suggested this treatise was "une espèce de profession de foi de l'École, moyennant quoi l'École d'Alexandrie traverse sans dommages la tourmente qui emporta l'École d'Athènes." In the concessions that Ammonius made to the Christians,

> l'homme vraiment providential fut un chrétien, Jean le grammarien, *philoponus* dans l'Église d'Alexandrie, qui couvrit son maître en publiant sous son nom accolé au sien ses immenses commentaires sur Aristote et en rédigeant au moment opportun une réfutation en règle de Proclus, qui désolidarisait définitivement le sort de l'École d'Athènes et celui de l'École d'Alexandrie.[17]

Philoponus may have written the treatise as a 'profession of faith,' but it is more likely that he wrote it on behalf of the Monophysite Christians than for a school of philosophy traditionally dominated by pagans. The timing of the appearance of Philoponus' treatise may be significant, since Justinian had just recently become emperor. Philoponus may have written his *De Aeternitate Mundi* to demonstrate, at an important moment, the Christian fervor of the Monophysites on an issue of common orthodoxy. The Monophysite controversy was a far more important problem, for the Emperor as well as the Alexandrians, than the presence of a school of pagan philosophy in Alexandria. As an hypothesis, it makes more sense to relate the publication in 529 of Philoponus' *Contra Proclum* to this religious controversy than to the survival of Ammonius' school.

In summary, the survival of the school of pagan philosophy at Alexandria may not have been a result of a decision on the part of the imperial authorities, but rather the result of negligence due to preoccupation with more pressing affairs. Furthermore, the publication of Philoponus' diatribe against Proclus undoubtedly had more to do with the religious schism that absorbed Egyptian Monophysites at this time than with the fortunes of the Alexandrian school of pagan philosophy.

Leaving aside for the moment the purely historical considerations surrounding the survival of Ammonius' school, let us evaluate what is known from the commentaries alone about Philoponus' relation to Ammonius' school. Our most reliable indication of the fidelity of Philoponus' commentaries to the lectures of his teacher is the evidence taken from the titles of those commentaries. These titles are also the best indication of his status in the school. Marcel Richard wrote the classic article on the various sorts of Byzantine commentaries as indicated by their titles.[18] Based on his analysis, certain conclusions can be drawn from the titles of Philoponus' surviving commentaries, listed below.[19]

Categories	*scholia eis tas Katêgorias*
Prior Analytics	*scholikai aposêmeiôseis ek tôn sunousiôn Ammôniou tou Hermeiou*
Posterior Analytics	*scholikai aposêmeiôseis ek tôn sunousiôn Ammôniou Hermeiou meta tinôn idiôn epistaseôn*
De Anima	*scholikai aposêmeiôseis ek tôn sunousiôn Ammôniou Hermeiou meta tinôn idiôn epistaseôn*
Physics	no descriptive term for the commentary as a whole
On Generation And Corruption	*scholikai aposêmeiôseis ek tôn sunousiôn Ammôniou Hermeiou meta tinôn idiôn epistaseôn*
Meteorology	no descriptive term for the commentary as a whole

Almost all of these titles refer to the commentaries as *scholia* or *scholikai aposêmeiôseis*. Richard pointed out that *scholia* probably evolved from *scholia apo phônês*, the latter term meaning notes taken from an oral presentation. He thought that Ammonius probably designated a number of different students to record his lectures. This accounts for the existence of commentaries on the *Categories* and *Prior Analytics* taken *apo phônês Ammoniou* without any indication of the recorder's identity. The existence of commentaries under the names of Philoponus and Asclepius indicate that each of these students made some original contribution to his own redaction. It appears that Philoponus exercised even greater liberty in his redaction, because four of his commentaries (on *De Anima*, the *Prior* and *Posterior Analytics*, and *On Generation and Corruption*) are taken *ek tôn sunousiôn Ammoniou*, rather than *apo phônês*, like Asclepius' commentary on the *Metaphysics*. This judgement is strengthened by noting the addition, *meta tôn idiôn epistaseôn*, to three of these titles.[20] Based on this information, the most reasonable assessment would be that Philoponus was one of a number of students authorized at different times to report Ammonius' lectures. He made some contribution of his own to the redaction, which was an acceptable practice. He may have made more changes than was considered prudent, however. His deviation from the conservative task of *reportage* may be what led Simplicius to accuse him of plagiarism, as if he claimed as his own traditional ideas which represented, in large part, the thought of his predecessors.[21]

From this evaluation of the degree of originality in Philoponus' commentaries, we can establish the chronology of Philoponus' works and trace his intellectual development. Gudeman thought that all the commentaries belonged to Philoponus' early life, in view of the absence of Christian doctrine in them. He thought Philoponus converted to Christianity sometime between 517, the date of the commentary on the *Physics*, and 529, the date of *Contra Proclum*, and that he only wrote theological works or refutations of pagan philosophy after his conversion.[22] Scholars now agree that the absence of Christian references is inconclusive as proof of Philoponus' religious conviction at the time of writing the commentaries. In the first place, the author's own opinion need not impinge on the explication of the meaning of a pagan text, especially since much of it represented a scholarly evaluation handed down through the schools. To the extent that Philoponus criticized Aristotle and presented his own solutions to philosophical problems, when his comments are relevant to Christian doctrine, they are such as to allow latitude for an interpretation consistent with Christian doctrine.[23] The chronology which is accepted here is based on the supposition that Philoponus was never a pagan.[24] It places the commentaries on *De Anima* and the *Categories* before 517, the date of the commentary on the *Physics*. The commentary on the *Prior Analytics* probably belongs to this early period as well. The commentaries on the *Physics* and *Posterior Analytics* belong to a middle period dating before 529. The commentary on the *Meteorology* appeared after 529, the date of *Contra Proclum*. The *Contra Aristotelem*[25] appeared after 529 but before 560, the terminus post quem for *De Opificio Mundi*.

Verrycken has recently argued for the presence of two different philosophical systems in Philoponus' intellectual career taken as a whole. The first system is Alexandrian Neoplatonism and comprises the teachings of Ammonius. The second system, represented by the theological works written after 529, is not Neoplatonic but deistic and theological. Between the two periods, Verrycken sees the two strata as mixed. In particular, he thinks Philoponus reworked parts of the commentaries on the *Physics* and *Posterior Analytics*, though each existed before 529. At present, it appears that Philoponus evolved a philosophical viewpoint of his own, starting from one closer to the Neoplatonism represented by the tradition of commentary on *De Anima*. However, even at this early stage, his opinions were not identical with the Neoplatonism of his predecessors.[26] To what extent Philoponus evolved not just a theological, but a philosophical standpoint clearly independent of Neoplatonist metaphysics is an important question. In the course of presenting Philoponus' ideas in the following pages, I will from time to time compare his ideas to those of Proclus or Simplicius to bring evidence to bear on this larger question.

Notes

1. John Malalas, *Chronographia; ex recensione Ludovici Dindorfii* (Bonn: E. Weber, 1831), 451.

2. See Alan Cameron, "The Last Days of the Academy at Athens," *Proceedings of the Cambridge Philological Society*, n.s.15 (1969): 15-17, and L.G. Westerink, ed., *Anonymous Prolegomena to Platonic Philosophy* (Amsterdam: North Holland Publishing Co., 1962), xviii.

3. Justinian's revised code of 529 included a law forbidding pagans to teach. This was only one of many laws added to the code which increased the repressive measures against pagans and non-orthodox Christians. Pagans could not inherit property or will property to the non-orthodox, nor could they enter into litigation against orthodox Christians. Pagans could not serve in the military or receive any public honor (*Codex Justinianus*, ed. by Paul Krueger, I.5.18). These proscriptions were reinforced by a restriction against pagans enjoying compensation from the public purse. Also, pagans who taught were threatened with confiscation of property and exile (*Codex*, I.10.2, I.11.9-10). This prohibition applied not only to the teaching of the doctrines of pagan religions and doctrines of philosophy inconsistent with Christian religion. It specified that no doctrine could be taught by pagans. Presumably then, pagans could not even teach grammar or rhetoric, disciplines which by the sixth century were thoroughly acceptable to Christians. In light of the general prohibitions against paganism, a special edict directed at the Academy hardly seems necessary.

Bury suggested that the order Malalas described was not a special edict, but rather a rescript, sent by Justinian in response to a request from the proconsul for instructions on how to interpret the law:

> Justinian had already taken stringent measures against pagans.... It is not difficult to guess what happened. The edicts against paganism, strictly interpreted, involved the cessation of Neoplatonic propagandism at Athens. The schools went on as before, and in a month or two the proconsul for Achaia would communicate with the Emperor on the subject and ask his pleasure. The *prostaxis* mentioned by Malalas was the rescript to the proconsul (Edward Gibbon, *The History of the Decline*

and Fall of the Roman Empire, ed. by J.B. Bury, 2nd ed. (London: Methuen, 1901), vol.4, 266, note 154).

If Bury's suggestion is correct, this rescript was a straightforward reiteration of already existing laws that applied to the Neoplatonic Academy.

It is possible, however, that the *prostaxis* was completely independent of the laws against pagans. Hugh Mason identified *prostaxis* as one Greek term for *mandatum*, imperial instructions to a subordinate official (*Greek Terms for Roman Institutions: A Lexicon and Analysis* (*American Studies in Papyrology* 13 (1974): 81). These instructions were made upon the initiative of the emperor and not in response to a specific request. If Justinian's order was a *mandatum* rather than a *rescriptum*, the closing of the Academy may reflect the emperor's will on an issue unrelated to the laws against pagans. It could have been directed against the practice of philosophy and law in Athens, rather than the presence of pagans in those schools. As such, it could reflect Justinian's desire to centralize higher education in Constantinople. From this standpoint, Malalas' report of the closing of a law school in Athens may be related to Justinian's decision, a few years later, to close all the law schools except those at Constantinople, Rome, and Berytus. There are reports that Justinian revoked the public salaries of teachers throughout the empire (Procopius, *Anecdota* 26). On the decline of teaching and consequent decline in general culture under Justinian, see Paul Lemerle, *Le premier humanisme Byzantin: notes et remarques sur enseignment et cultur à Byzance des origines au X*[e] *siècle* (Paris: Presses universitaires de France, 1971), 70-73. We cannot rule out the possibility that Justinian wished to discourage higher education through the Empire and to encourage only a small number of intellectuals to establish themselves in Constantinople.

4. Olympiodorus, *In Alcibiadem*, ed. L.G. Westerink (Amsterdam: North Holland Publishing Co, 1956), 92: 141.3 On the dating of this passage, see Westerink, ed., *Anonymous Prolegomena*, xiv-xv, and Cameron, "Academy at Athens," 11-12.

5. Westerink, ed., *Anonymous Prolegomena*, xiii, xx.

6. See E.R. Dodds, *The Greeks and the Irrational* (Berkeley: University of California Press, 1951), Appendix II, 283-311.

7. Damascius, *Reliquiae Vita Isidori*, ed. and annotated by C. Zintzen (Hildesheim: Georg Olms Verlagsbuchhandlung, 1967), 79, p.110.

8. See Philip Merlan, "Ammonius Hermiae, Zacharias Scholasticus and Boethius," *Greek, Roman, and Byzantine Studies* 9, no.2 (1968): 193-203 on Zacharias' attitude toward pagan philosophy, and Averil Cameron, *Agathias* (Oxford: Clarendon Press, 1970), 102-4, 112-15, for a discussion of Agathias' attitudes toward pagan natural philosophy.

9. Damascius, *Isid.* fr.316 (Zintzen, 251).

10. Paul Tannery, "Sur la période finale de la philosophie grecque," in *Mémoires scientifiques* VII, 225. We can be sure that the Alexandrian school in the sixth century enjoyed no private income, as did the Neoplatonic Academy. The municipal chair was undoubtedly a salaried post, at least when originally instituted. However, throughout the Empire, the greater part of a salaried teacher's income derived from student fees (A.H.M. Jones, *The Later Roman Empire, 284-602* (Oxford: B. Blackwell, 1964), vol.2, 1002), and there is no reason to think the situation was different in Alexandria. Remarks by Olympiodorus indicate that student contributions formed part of the income of the philosophers at Alexandria (*In Alcibiadem*, 140.18-141.1 (Westerink, 92); see also Westerink's article, "Philosophy and Medicine in Late Antiquity," *Janus* 51 (1964): 176-77. Consequently, from a strictly financial point of view, the local or imperial authorities may have had little power to stop the teaching of the *philosophos* as long as students were willing to pay for the privilege of hearing his lectures. However, the dependence of the Alexandrian *philosophos* on student fees entailed financial insecurity and dependence on the good will of well-to-do students. This was a possible source of coercion of which the Athenian school had been independent.

11. H.D. Saffrey, "Le chrétien Jean Philopon et la survivance de l'École d'Alexandrie au VIe siècle," *Revue des études grecques* 67 (1954): 401.

12. Pierre Courcelle, *Les Lettres grecques en Occident* (Paris: E. de Boccard, 1943), 296.

13. J.P. Migne, ed. *Patrologiae Cursus Completus, series Graeca* LXXXV, 1029-1117.

14. W.H.C. Frend, *The Rise of the Monophysite Movement* (Cambridge: Cambridge University Press, 1972), 241-42.

15. A.A. Vasiliev, *Justin the First* (Cambridge: Harvard University Press, 1950), 119.

16. This political power consisted primarily in the power to disrupt, by inciting the Christian folk to violence. For an account of the history of non-cooperation in late antique Egypt, see Norman H. Baynes, "Alexandria and Constantinople: A Study in Ecclesiastical Diplomacy," *Journal of Egyptian Archeology* 12 (1926): 145-156.

17. Saffrey, "Le chrétien Jean Philopon," 408.

18. Marcel Richard, "*Apo Phônês,*" *Byzantion* 20 (1950): 191-222. For a different approach, see L.G. Westerink, "Deux commentaires sur Nicomaque: Asclépius et Jean Philopon," *Revue des études grecques* 77 (1964): 533-34. Westerink suggested that Philoponus may have been excluded from Ammonius' lectures and that he had neither the approval nor permission to publish them. Evrard criticized Westerink's suggestion in "Jean Philopon, son commentaire sur Nicomaque et ses rapports avec Ammonius, "*Revue des études grecques* 78 (1965): 596-97.

19. In some cases, there is no title for the whole work separate from the title to the commentary on the first book. In these cases, the titles to subsequent sections represent a shorter form of the title of the first section. The commentary on *Generation of Animals* bearing Philoponus' name has been excluded from his list, because it was probably written by Michael of Ephesus. See the editor's preface to that commentary (vol. 14, part 3, in *Commentaria in Aristotelem Graeca*, ed. M. Hayduck (Berlin: G. Reimer, 1882-1909).

For a complete list of all the known works of Philoponus, those that survive as well as those that do not, together with works of doubtful attribution, see A. Gudeman's article on Philoponus, *Real-Encyclopädie der klassischen Altertumswissenchaft* IX, 1772-95. Philoponus' theological works have been excluded from consideration in this introduction, with the exception of his two polemics against the eternity of the world and his *De Opificio Mundi*. All three are important for tracing the development of Philoponus' philosophical ideas, relating them to his Christian convictions, and also for dating the commentaries, since the dates of his *Contra Proclum* and *De Opificio Mundi* are known from internal references.

20. Richard, "*Apo Phônês*," 192-94.

21. Simplicius accused Philoponus of appropriating as his own one of Xenarchus' objections to Aristotle's idea of circular motion (Simplicius, *In De Caelo* (vol. 2 in *Comm. in Arist. Gr.*, ed. by I.L.Heiberg), 42.17-20).

22. Gudeman, *Real-Encyclopädie*, 1769.

23. Étienne Evrard, "Les convictions religieuses de Jean Philopon et la date de son commentaire aux «Meteorologiques»," *Bulletin de l'Academie Royale de Belgique*, Classe des Lettres 39 (1953): 349-355. Part of the problem in evaluating the commentaries for evidence of Philoponus' religious conviction is that it is not clear what doctrines were acceptable in Christian circles at this time. On this point, Michael Wolff criticized Evrard for suggesting Philoponus held an Originist view of the eternality of souls. See Wolff's *Fallgesetz und Massebegriff: Zwei wissenschaftshistorische Untersuchungen zur Kosmologie des Johannes Philoponus* (Berlin: Walter de Gruyter, 1971), 80.

24. This chronology concerns only Philoponus' surviving philosophical works. It results from Evrard's efforts to establish the development of Philoponus' thinking on the nature of the circular motion of the heavens and sublunary sphere (Evrard, "Les convictions religieuses," 305-22), and from Verrycken's separation of two strata of philosophical thinking in Philoponus' middle commentaries ("The development of Philoponus' thought and its chronology," *Aristotle Transformed*, ed. by Richard Sorabji (Ithaca, N.Y.: Cornell University Press, 1990), 233-274. Wolff confirmed Evrard's placing of the commentary on *De Anima* before 517 and suggested as well that the commentary on the *Categories* was written before 517 (Wolff, *Fallgesetz und Massebegriff*, 6, 120). The place of the commentary on *On Generation and Corruption* in this chronology was not determined by Evrard. Gudeman placed it after the *Physics* commentary but before the commentary on the *Meteorology* (Gudeman, *Real-Encyclopädie*, 1780). Verrycken placed it among Philoponus' earliest comentaries. Neither Evrard or Wolff attempted to date the commentaries on the *Prior* and *Posterior Analytics*. Following the ideal curriculum of studies outlined by many ancient commentators,

Gudeman supposed that the logical works comprise the first group of commentaries written by Philoponus. However, Verrycken pointed to passages in the *Posterior Analytics* commentary that reflect Philoponus' later point of view. My research in Chapters Six and Seven connecting the commentaries on the *Physics* and *Posterior Analytics* tends to confirm Verrycken's judgement on this point.

25. *Contra Aristotelem* survives only in fragments, most of them preserved by Simplicius in his commentary on *De Caelo*. Evrard collected the fragments in his unpublished thesis "Philopon, Contre Aristote, Livre I" (Université de Liège, 1942-43). More recently, see *Philoponus: Against Aristotle on the Eternity of the World*, trans. by Christian Wildberg (London: Duckworth, 1987).

26. For the relation of Philoponus' commentary to other commentaries on *De Anima*, see Wolff, *Fallgesetz und Massebegriff*, 70-72, and H.J. Blumenthal, "Neoplatonic Elements in the *De Anima* Commentaries," *Phronesis* 21, no.1 (1976): 64-87, especially 84-86.

Chapter Two

Reconciling the Mathematical and Physical Aspects of Optics

The history of optics in antiquity consisted of two unreconciled elements: geometrical optics and philosophical ideas about light. Geometrical optics, exemplified in the treatises of Euclid and Ptolemy, was the mathematical treatment of perspective and reflection, based on the hypothesis of imaginary lines originating in the eye and terminating at the visible object. However, it was in the context of philosophical discussions about sense perception that questions arose concerning the nature of light. In these discussions, light was defined in terms appropriate to the explanation of sensation. The history of optics in late antiquity was dominated by a physical theory of light and vision based on the emission of visual rays from the eye, simply because it was difficult to combine the basic rules of geometrical optics with any theory of vision that denied the emanation of visual rays from the eye.[1] John Philoponus made just this denial in proposing his own reconciliation of geometrical optics with Aristotle's theory of light and vision. Philoponus' theory for the propagation of light and color was the cornerstone of this reconciliation. This chapter will show the role of Philoponus' theory in combining geometrical optics with a theory of vision in which visual rays play no substantial part.

First, let us examine briefly the basic principles of geometrical optics. Euclid, who introduced the mathematical science of perspective in the third century B.C., made no claims about the physical nature of the phenomena he described mathematically. His proofs were based on the rectilinear propagation of visual rays from the eye. Diverging, the rays form a cone, whose base lies at the visible object and whose axis at a point on the pupil.

The apparent size of a body is determined by the angle formed by the visual rays at the object. Since the rays diverge as they emanate from the eye, the visual cone is not continuous, and once the rays strike a body, they cannot perceive it in its continuity. As a result, rapid movement of the visual cone over the body is necessary in order to perceive the object in its entirety. Consequently, objects seen under more angles appear more distinctly.[2] Euclid used both the terms *opseis* and *aktines* to refer to visual rays, but in proof of one of his propositions, he also used *aktis* in reference to a ray of light from the sun. In this proof, which involved measuring the height of an object by the length of its shadow, he assumed that a light ray, like a visual ray, is propagated along a straight line.[3] Nevertheless, Euclid articulated no presuppositions about the physical nature of visual rays or light rays.

Theon of Alexandria, lecturing on Euclid in the fourth century, explicitly stated that propagation along straight lines is a trait shared by light rays and visual rays. However, he accorded to light rays no role in vision. Furthermore, he argued against those who believed visual perception occurs due to something emitted from visible objects and received in the eye. He said that if vision resulted from an influx from the visible object, there would be no reason why one should fail to see something in his field of vision. Everyone knows it is possible to overlook a small object. This is due, Theon believed, to the discontinuous nature of the visual cone. Visual rays do not strike all points within the visual field and consequently, small objects may be overlooked.[4] His designation of both visual rays and light rays has indicated to some that he, as well as Euclid, shared Plato's ideas about visual perception articulated in the *Timaeus*.[5] Plato argued that fire is emitted along the line of vision from the eye. The light outside coalesces with this fiery flow to permit vision. Without light outside, the flow from the eye is extinguished and this is why we cannot see in darkness. While Theon recognized both light rays and visual rays, he did not speak of the visual cone as fiery in nature, nor did he describe any combination of visual rays and light as necessary for vision.

Ptolemy, author of the second great treatise on geometrical optics, took a greater interest in explaining the nature of light and vision. The first book of Ptolemy's *Optics* is lost to us (it was, in all probability, also lost to the Arab translators of Ptolemy in the ninth century). This is unfortunate, since Ptolemy's remarks at the beginning of the second book indicate that the first book was specifically concerned with the nature of light and vision.[6] The part of the *Optics* which survives is primarily mathematical in character, but there are scattered indications of the content of the first book. Albert

Lejeune has assembled these passages, which indicate, among other things, an opinion about the role of light in vision.[7] Like Aristotle, Ptolemy believed that color is the proper object of vision and that spatial characteristics are perceived in a secondary way. Light and color are closely related, both being species of the luminous, but the presence of light is a prerequisite for the perception of colors. Ptolemy believed that the cooperation of light is necessary for visual rays to perceive color, but lacking the first book of his *Optics*, we do not know what he considered the nature of the cooperation between light and visual rays, or even if he addressed this issue.

Ideas about the nature of light developed outside of geometrical optics. Light was treated as a peripheral issue in the larger problem of sense perception. Aristotle's treatment of light, in *De Anima* II.5-7, subordinates the question of the nature of light to the description of the physical process of perception. In choosing a model for perception, Aristotle took into account the views of his predecessors, some of whom maintained that perception occurs by the interaction of like elements, and others, by unlike elements. One problem of these earlier theories was that they did not distinguish between perception and ordinary physical interaction. Consequently, there was no reason why the senses perceive at one time and not at another. For example, Aristotle criticized the like-like theorists by saying they gave no reason why the senses are not continually perceiving themselves (*De Anima* II.5, 417a2-6). Aristotle himself solved this problem by claiming that the ability to perceive always exists potentially in the organ of perception, but it is led into actuality only by the presence of an appropriate object of perception, acting as an agent cause. One possesses sense perception in the same way a person always has knowledge of a subject in which he possesses some expertise, even when he is not exercising that knowledge. Perception may exist potentially or actually, depending on whether or not a visible object is present to activate perception. Perception is potentially like its agent cause, but unlike it until the potential is made actual.

Each of the senses is actualized, when acted upon by a perceptible object specific to that sense. The proper object of vision is color. However, for color to be visible, something else is required--namely, the presence of light. Aristotle treated light solely in relation to its role in visual perception. Consequently, it is not surprising that his description of light has as its model the doctrine of perception. He defined light as the actuality of a trait that always exists potentially in the medium. This trait is transparency. The color of a visible object sets up a movement in the transparent medium which, in turn, affects the eye. The visual faculty is actualized and the object is seen.

Aristotle was not the first to explain vision on the basis of an effect of the visible object received, directly or indirectly, in the eye. However, Aristotle's theory had a striking advantage over rivals--an advantage which Aristotle was quick to point out. Light is not a body, but an attribute of the medium, which is itself a body. Consequently, it is possible for light to exist in the medium without there being two bodies in the same place at the same time. Those who believed that light is a species of fire encountered considerable problems in explaining how light is propagated from a luminous object to a point across a room--or in the case of the heavenly bodies, across the universe. This advantage of Aristotle's theory was an important component of Philoponus' refutation of the visual ray hypothesis. He argued against the idea that anything is emitted from the eye in vision. He claimed that any such emission must be corporeal. As such, it would encounter all the difficulties of one body passing through another. Many of his arguments against corporeality are the same for vision and light. These arguments, which Philoponus repeated for light after first applying them to vision, did not originate with Philoponus, but have a long history in the Aristotelian commentary tradition. Similar arguments were used by Alexander of Aphrodisias against the visual ray hypothesis.[8]

If visual rays are corporeal, Philoponus asked, how do these rays traverse the medium? Either they pass through the medium itself or they cut it. The former alternative is absurd, since body cannot pass through body. But, by the latter alternative, we would be unable to see anything in its continuity. For we would not see the parts of the visible objects behind the parts of the air which are not cut. As a result, we would see everything divided into parts (*In de An.*, 325.6-33). He used the same types of argument against the exit of visual rays from the eye. If the rays cut the pupil, then its fluid would flow out, but it is impossible that body should pass through body. It could be that the eye has pores, through which the rays take their exit, but this possibility meets with the same problems as the passage of rays through a medium by cutting it. The rays could pass outside only where there existed pores and not where there were particles separating the pores. As a result, we would be unable to see things in their continuity (*In de An.*, 326.15-26). When Philoponus went on to argue that light is not a body, the same alternatives appeared. If light is a body, either it passes through the whole air, or cuts it, or passes through the pores of the air. The first alternative is impossible, since body cannot exist in body. The other choices he rejected as a pair. If light either cuts the air or passes through its pores, then light will be in some places and air in other places. In other words, only certain parts of the medium will be illuminated. Furthermore, there will be times when we breathe light only and not air. Also, the wind ought to darken the place where it blows, displacing the light. But all these consequences are absurd (*In de An.*, 327.25-39).

These recurrent types represent arguments found in a more extended form in Alexander of Aphrodisias' *De Anima libri Mantissa*. Alexander also wondered how rays can cross the medium. Body cannot pass through body, but if the medium has pores, either there exists a separable void or else the pores are filled with air or some other body. In the latter case, the corporeal rays again must pass through body. If the medium has pores, the transparent should not appear continuous to us but rather perforated, like a sieve. And visible objects should be mottled with shadow (*Mantissa*, 129.25-32). Like Philoponus, Alexander used similar arguments in relation to the pupil (*Mantissa*, 128.24-129.20). If visual rays exit from the pupil, either they do so in a continuous cone, or divided into parts. If as a continuous cone, then the pupil itself must be one large pore, which is not the case. If the rays exit separated from one another, they may coalesce outside in the medium, or remain separated as they extend from the eye. In the latter case, a great deal that lies within the visual cone will not be seen, due to the distance between the visual rays. When Alexander turned to argue that light is not a body, the same alternatives arose: either light passes through the medium, or the medium has pores (*Mantissa*, 139.1-17). He explored the problems involved in saying these pores are void or filled with some sort of body. One of these problems was also discussed by Philoponus: where there are pores, the medium would be illuminated, but where there are no pores, it would be dark. Philoponus' arguments against the corporeality of light and visual rays are not as lengthy or complex as those of Alexander. His arguments are of interest, however, because they represent the application to the case of light of an exegetical pattern common among the Aristotelian commentators.

This pattern involved the development of all the problems implied by two bodies occupying the same place at the same time. Aristotle's treatment of light is only one of the passages where he used this impossibility to refute a viewpoint contrary to his own. In the commentary tradition, this concern about two things occupying the same place at the same time became the problem of body passing through body, and where Aristotle evoked the impossibility of interpenetration, the commentators deployed the arguments against body passing through body.[9] The arguments cluster around two issues: either bodies occupy the same place or one has pores into which the other diffuses itself. Alexander refuted both alternatives, considering all the possible forms each might take. For Philoponus, however, the broad outlines of the issues sufficed to evoke what were, by his day, conventional arguments.

Both Alexander and Philoponus thought that if vision occurred by an emission from the eye, the emission must be of a corporeal nature. Alexander stated this flatly at the beginning of his own refutation of the visual ray hypothesis (*Mantissa*, 127.28-29). Philoponus, however, offered one argument against the possibility that the emission from the eye is incorporeal (*In de An.*, 326.26-37). An incorporeal entity must have a vehicle (*oxêma*), or material substratum.[10] If the incorporeal ray passes out of the eye with the aid of its own vehicle, then it passes out as a body, and all the previous arguments apply. If separated from its own substratum, it must make use of the medium. The problem with this latter alternative is that the incorporeal ray is a power capable of perceiving (*aithêtikê dunamis*), and therefore, it is alive. This living power, coming to be in the medium, would make the medium itself alive and capable of perception, or else the living would exist in the non-living. The first alternative is ridiculous, and the second impossible. Consequently, visual rays could not be incorporeal. By this argument, Philoponus made it clear that anyone who held to the visual ray hypothesis must consider those rays corporeal. He made the visual ray theory subject to all the pitfalls of explaining how a body passes through a material medium. He reserved for his own intromission theory all the advantages of an incorporeal entity for explaining transmission. The most important aspect of the incorporeality of light is that it can be received and transmitted by other bodies.

In his general exegesis of *De Anima* II.7, 418b9 (*In de An.*, 324.23-35), Philoponus paraphrased Aristotle's doctrine of light in a simple and straightforward fashion. Even in this introductory segment, there is a subtle shift of meaning in the definition of light. Aristotle had considered the actuality of transparency and light identical. Philoponus considered light as something which a medium can receive and transmit, if it possesses the trait of transparency. When light supervenes (*epigignomenon*), then it is possible to see through a transparent medium.[11] Ranging light with color and separating it from transparency, he now defined transparency in terms of its ability to act as a carrier for light and color. The transparent has a power for conveying color, although it is itself colorless. The air is completely unaffected by the colors that pass through it, and this is the reason why viewers looking in different directions through the same portion of air all see different objects without interference or distortion; the same portion of the air can convey any number of colors at the same time without being affected by any of them. Colors make use of the medium as a substratum and act on vision by means of the air. Philoponus compared the action of color on vision to the chopping of wood by a carpenter. The force of his blow passes through the axe and shapes the wood, while the axe itself, like the air, remains unaffected. Light is received and transmitted by the air in the same way:

> So, just as in these cases, so also in the case of light, we say that a certain incorporeal actuality is sent out of a body capable of illuminating into transparent things which naturally receive this (*In de An.*, 329.35-37).[12]

Philoponus compared the transmission of light and color to the eye to the way in which fire warms us, even when it lies some distance away. The fire does not come up to us, but rather it warms the air nearest itself, and this air in turn warms the air next to it, which also passes on the effect, until things very distant from the fire receive the effect and are themselves warmed. Philoponus said that the same thing happens in the case of light:

> ... the sun acts directly in the body neighboring it and in contact with it and makes it actually transparent, and that [body], having become such, possesses the power so that it also acts on the things nearby and makes them such, and in this way, the illuminating power (*phôtistikê dunamis*) proceeds out to the limits, not by the actuality of the sun proceeding with no intermediate to the limits, but by those nearby that are acted on primarily being able themselves to actualize the similar actuality[13] in those capable of receiving it (*In de An.*, 330.8-14).

According to this explanation, light is propagated, not by a movement from place to place, but by a succession of effects. Once it has received the effect, each member of the succession possesses the power to affect its neighbor in a similar way. Philoponus went on to claim that this transmission takes place all at once and without the passage of time (*In de An.*, 330.14-15). To explain this, he invoked only the incorporeality of the actuality transmitted, and he justified it with another simile (15-19). If someone moves a long rope by grasping the tip, the whole rope will move along with the tip without the passage of time, due to the continuity of the parts of the rope. This is also how light from the sun is propagated all at once, because all the parts of the universe are successive.

This theory for the propagation of light and color is the vehicle for Philoponus' reconciliation of geometrical optics with an intromission theory of vision. Incorporating geometrical optics into his own version of Aristotle's theory of light, he also solved one of the major objections to Aristotle's theory. According to Aristotle, every part of the transparent should be filled with the colors of visible objects from every direction, and we should be able to see even the things behind us. To solve this problem, Philoponus said he would make use of the same supposition the visual ray theorists use:

> So, just as those [philosophers] suppose that visual rays and [light] rays are sent out along straight lines and are reflected from smooth bodies at equal angles, in this way also we suppose that actualities of colors, and furthermore, even light, are sent out along straight lines and reflected from smooth bodies at equal angles. So for this reason also, images in mirrors appear, not because our visual rays are reflected at the visible objects, but because their actualities come to us (*In de An.*, 331.3-10).

Light and color proceed along straight lines from their sources, but we see only those visible objects that lie on straight lines with the eye. This is why we do not see the things behind us, even though their actualities are in the air (*In de An.*, 331.14-15). Philoponus extended the explanation to reflection, saying our visual rays are not reflected from mirrors, but rather light and color travelling from visible objects are reflected and proceed along straight lines from the mirror to the eye. He asserted that all the basic principles of the visual ray theory can be assimilated to an Aristotelian viewpoint:

> And to put it simply, whatever they say concerning visual rays, we say the same things in precisely similar terms concerning actualities and save the appearances. For what difference does it make whether straight lines come to the mirror proceeding from the eye or are reflected from the mirror to the eye? Then, if these are held in common by those [philosophers] and the doctrine of Aristotle,

> but the one in terms of actualities, the others in terms of visual rays, and myriad impossibilities follow from the supposition of visual rays, one must rather choose the supposition of Aristotle, which both saves the appearances and avoids the impossibilities (*In de An.*, 331.25-33).

Philoponus devoted the greater part of the commentary on 418b9 to demonstrating that his version of the Aristotelian hypothesis does indeed preserve the appearances, while eliminating the absurdities of other theories. The commentary on 418b9 aims to produce a thorough and consistent intromission theory based on Aristotle's brief remarks about light. Although Alexander shared Philoponus' concern with the refutation of the visual ray hypothesis in favor of Aristotle's theory of light, his expectations did not extend to explaining the propagation of light or reconciling geometrical optics to an intromission theory of vision. For him, it was enough that light is actualized throughout the medium all at once, and that the color from objects affects the eye by means of the medium. Philoponus' theory for the propagation of light does not belong to the Aristotelian commentary tradition. Rather, it reflects his own will to make Aristotle's theory of light consistent with the basic principles of geometrical optics, and thus to remove these principles from their habitual association with the visual ray hypothesis.

The theory for the propagation of light is the means by which Philoponus sought to reconcile the mathematical and philosophical aspects of optics. However, there are separate problems connected with Philoponus' theory for the propagation of light and his remarks about the compatibility of the mathematical and philosophical aspects of optics. In this book, a central problem connected with each part of Philoponus' treatment of light will be considered.

The first problem concerns the propagation of light and originates in Philoponus' statement that light is propagated all at once or, as he puts it in more precise terms, without the passage of time (*achronôs*) (*In de An.* 325.6-8). For Philoponus, it was a readily observable fact that both light and color are present throughout the medium, as soon as a light source is present. A luminous body communicates an incorporeal effect to the medium, which in turn acts on the visual faculty. Even though the effect spreads throughout the medium by a succession of effects, nevertheless this succession is accomplished without the passage of time. From a strictly Aristotelian point of view, however, it is unacceptable that successive change should ever be accomplished without the passage of time, regardless of whether that succession is the change of place of a body or the transmission of an incorporeal effect. The exposition of this problem and its resolution are presented in Chapter Three. Other issues raised during the course of that examination lead to an investigation of the causal theory common to Philoponus' theory for the propagation of light and his impetus theory. This investigation, centering on Philoponus' commentary on *Physics* III.3, is the subject of Chapter Four. Chapter Five shows how the causal theory developed by the commentators, in the interpretation of *Physics* III.3, was also used in the formulation of Neoplatonic metaphysics. This chapter suggests that there was already a strong Aristotelian component in the participation theory thought by many to have influenced Philoponus. Taken together, the issues treated in these chapters reveal the philosophical origins of the idea of imparted power which is the essential component in Philoponus' theory for the propagation of light.

The second problem treated in this book concerns the reconciliation of the mathematical and physical aspects of optics, and it originates in Philoponus' idea that it makes no difference, from the point of view of geometrical optics, whether rays proceed from the visible object to the eye, or from the eye to the visible object (*In de An.*, 331.27-29). Chapters Six and Seven investigate how Philoponus' view of the nature of mathematical science lends support to this statement. Chapter Six shows that Philoponus changed the Aristotelian approach to defining the subject matters of optics and astronomy, by granting to three-dimensional continuity a kind of existence prior even to substance in the order of nature. Chapter Seven explores the implications of this move for Philoponus' understanding of the relation of the mixed sciences--optics, astronomy, and harmonics--to mathematics.

The examination of both these topics--Philoponus' theory for the propagation of light and his hypostasizing of three-dimensionality--shows that Philoponus' fundamental concerns were different from Aristotle's. In his treatment of both light and the subject matter of mathematics, Aristotle sought a description of perception and thought that avoids logical mistakes and assumes no extra existents. In his theory for the propagation of light, Philoponus appears to have made the fundamental mistake of thinking a succession need not take time. And his hypostasizing of three-dimensionality sets up an extra entity. Nevertheless, Philoponus' thinking on both these topics developed from philosophical issues in Aristotle's text, issues that are not entirely resolved in the text and which, for that reason, remain important to modern scholars in ancient philosophy. In this book, Philoponus' thinking on each topic is viewed as a genuine philosophical development of issues implicit in Aristotle's own thinking.

Notes

1. David Lindberg has pointed out that there are assumptions present implicitly in Euclid's theorems which militate toward a physical interpretation of visual rays. See his *Theories of Vision from Al-Kindi to Kepler* (Chicago: University of Chicago Press, 1976), 13-14. For an analysis of the Stoic fragments on light and vision, see Robert B. Todd, "*Sunentasis* and the Stoic Theory of Perception,"*Gräzer Beitrage* 2 (1974): 251-61. For the influence of the visual ray hypothesis in late antique Stoicism, see H.G. Ingekamp, "Zur Stoischen Lehre von Sehen," *Rheinische Museum für Philologie* N.F. 114, no.3 (1971): 240-46.

2. Euclid, *Optica*, vol.7, *Euclidis Opera Omnia* , ed. by I.L. Heiberg and H. Menge (Leipzig: B.G. Teubner, 1895), def. I-VII, prop. I.

3. Euclid, *Optica*, prop. XVIII.

4. Theon of Alexandria, *Recensio Opticorum* (*Euclidis Opera Omnia*, vol.7). See Prologue, 144-54.

5. Plato, *Timaeus* 45b2-32. For the Platonic theory of vision ascribed to Euclid and Theon, see Paul ver Ecke, trans., *L'Optique et Catoptrique* (Paris: de Brouwer et Cie, 1938), xiii-xiv, xxvi-xxvii.

6. Ptolemy, *Optica* II.1, in *L'Optique de Claude Ptolémée dans la version latine d'après l'arabe de l'émir Eugène de Sicile*, ed. with commentary by Albert LeJeune (Louvain: Bibliothèque de l'Université, Bureaux du recueil, 1956), 11.

7. Albert Lejeune, *Euclide et Ptolemée*, (Louvain: Bibliothèque de l'Université, 1948), 15-83. Gerard Simon has argued that the locus of ancient optics does not lie in natural philosophy at all and, especially, that Ptolemy's *Optics* is not concerned with theory of light: "L'objet de l'optique de Ptolémée, c'est la vision, et non la lumière. Il est faux de considérer le rayon visuel comme un rayon lumineux inversé: la "parenté" du rayon visuel avec la lumière est à l'état liquide est "aqueux," ou tout solide "terrestre"("L'optique d'Alhazen et la tradition ptoléméenne: une nouvelle insertion dans le champ du savoir," in the unpublished proceedings of the

Dibner Institute conference, "Optics: The Transmission and Transformation of a Discipline," held May 31- June 2, 1990. His denial that, for Ptolemy, the visual ray can be regarded as a light ray in reverse is interesting, since Philoponus explicitly reversed the direction of the visual rays of geometrical optics and made the physical nature of these rays an issue. For more on Simon's interpretation of ancient optics, see his book, *Le regard, l'être, et l'apparence dans l'optique de l'antiquité* (Paris: Seuil, 1988).

8. For Philoponus, see *In de Anima*, vol.15 in *Commentaria in Aristotelem Graeca* XV, ed. M. Hayduck, (Berlin: G. Reimer, 1882-1909), 325.6-326.37; for Alexander, *De Anima libri Mantissa*, vol. 2, part 1 in *Supplementum Aristotelem*, ed. I. Bruns (Berlin: G. Reimer, 1885-1903), 127.27-130.13. On the authenticity of the *Mantissa*, see Robert B. Todd, *Alexander of Aphrodisias on Stoic Physics* (Leiden: E.J. Brill, 1976), 18 and n.86 on that page.

9. For an account of this problem among the commentators and its origin in a combination of passages from Aristotle, see Todd, *Alexander*, pp. 73-88. Richard Sorabji traced the late antique history of interpenetration, chemical combination, and mixture of bodies in his *Matter, Space, and Motion* (Ithaca, N.Y.: Cornell University Press, 1988), ch.5-7.

10. There is no need to assume that Philoponus meant anything more than a material carrier by *oxêma* in this context. Even if the pneuma is the vehicle of the perceptive power in the body (*Gen. An.* 736b30-34), only a material substratum would be available in the medium. The question is more complicated in relation to the immortality of the soul; see Proclus, *Elements of Theology*, ed. by E.R. Dodds, 2nd ed. (Oxford: Clarendon Press, 1963), Appendix II, 313-21.

11. This move away from the identification of light with the actuality of transparency is worth noting because of its importance to later explanations of the relation of light to the medium. The actuality of transparency survives in these later explanations primarily as the power of the medium to receive and transmit light. See, for instance, Alhazen's "Discourse on Light" (in the French translation by R. Rashed, "Le «Discours de la lumière» d'Ibn al'Haytham (Alhazen)," *Revue d'histoire des sciences et de leurs applications* 21 (Juillet-September 1968): 208-10).

12. All translations from the commentators are my own.

13. See Appendix I, note 5, for the meaning of *energein tên homoian energeian.*

Chapter Three

The Propagation of Light Without the Passage of Time

I

Philoponus maintained that light is present simultaneously throughout the medium, even though the method of propagation he assigned to light, a succession of effects, is a process which in Aristotelian terms always involves time. Aristotle himself was Philoponus' authority for the propagation of light in no time whatsoever. Consequently, it is imperative to inquire to what extent Philoponus was justified, on Aristotelian grounds, in claiming that light is propagated without the passage of time.

Philoponus never claimed that light is propagated instantaneously (*tôi nun*). He followed Aristotle in the belief that no change can take place in an instant.[1] It must always be possible to correlate stages of change with divisions of time. Instants are indivisible, and although they may mark the limits of time periods, they can never form together the continuum which is a time period. Progress from one thing to another always takes time, but the goal may be accomplished quickly or slowly--that is, in a period of time which is shorter or longer. This latter point yields one of Aristotle's arguments against motion in an instant. Let us say that in an instant, two bodies move different distances: the body moving faster covers the distance AB, and the slower, AC. The body moving faster will then cover the distance AC in a shorter time than it covers the longer distance, AB. As a result, the instant must be divisible. But instants are indivisible. Consequently, no motion can take place in an instant (*Physics*, VI.3,

234a24-31). The principle underlying this denial of motion in an instant is the divisibility of the continua of place and time in relation to motion. It is due to the possibility of correlating distance with longer and shorter periods of time that expressions of speed are possible in relation to motion. This is also the reason why it is possible to rule out instantaneous motion.

Another of Aristotle's arguments against change in an indivisible time strikes closer to the heart of his general analysis of motion. In *Physics* VIII.8, 263b9-26, Aristotle constructed a general case of change in the transition of an object, D, from non-white to white. In time period A, the object is non-white and in time period B, it is white. C is an instant, the limit between time periods A and B. What is the character of the object D at C? Does C belong to period A or B? There is no doubt that the change from non-white to white is complete at C, since that instant is the end of the period in which the object is non-white. As a result, either 1) D is white at C, or 2) having become white, it is still non-white, or 3) it is both white and non-white at the same time. He went on to say (b28-30) that if D is neither white nor non-white at C, but 'becoming white,' then it is necessary to postulate another indivisible to accommodate the change from 'becoming white' to 'white,' and an endless regress is set afoot. The only reasonable alternative is that D is white and only white at the indivisible time, C. D is 'becoming white' throughout a period of time preceding C. The moment at which the change is complete is the first moment of the object's being white. The upshot of the solution is that there is no period of 'becoming white' separable from the period of 'being non-white.' The change is taking place while the object is still in its previous state. It is possible for Aristotle to make this last point, because he has construed any change as a change from not-A to A. When one considers the same case as a change from black to white, then it is necessary to consider possible gradations in the course of the change. However, the simplest case reflects the basic elements of his analysis of motion. There is no time of 'becoming' apart from the preceding condition of the thing changing. When the thing 'has become,' it has achieved the object of its motion and is in the new state. The instant when this change is achieved is merely the first moment of the new condition. There can be no change in an instant, but all change takes place in time.[2]

Given Aristotle's rejection of change in an instant, his criticism, in *De Anima* II.7 and *De Sensu* VI, of Empedocles' opinion about the motion of light is puzzling. According to Aristotle, Empedocles was wrong in maintaining that light could travel all the way between the earth and its outer envelope without our noticing it:

> ... this is contrary both to the clear evidence of reason and to the appearances; [the movement] might escape our notice over a short distance, but that it escapes our notice over the distance from east to west is too great a claim (418b23-26).

This argument is based on the fact that light does not seem to be present near its source in the medium *before* it appears in the medium farther away from its source. Aristotle here affirmed that light seems to come to be (*gignesthai*) in all parts of the medium simultaneously, but denied that light traverses the medium in an imperceptibly short period of time. In making this criticism of Empedocles, Aristotle assumed the necessary connection between motion and time. That the motion of light takes time might go unnoticed over a short distance, but this could not be the case when the distance involved was extremely large. The same point was made in more precise terms in *De Sensu* VI, where Aristotle reported Empedocles as saying that light from the sun first reaches a midpoint before arriving at vision or at the earth:

> Empedocles, for example, says that the light from the sun arrives first in the intervening place before reaching the eye or the earth. It would seem, reasonably, that this occurs. For what moves is moved from one place to another, so that there must be some time in which it is moved from one place to the other. But every time is divisible, so that there would be a 'when' when the ray is not yet seen, but is moving in the intervening place (446a26-b3).

In this passage, Aristotle made explicit the relation between movement through the medium and time. He believed that sound and smell reach a midpoint before arriving at the organ of perception, but he rejected this model as an explanation for light proceeding from the light source to the eye. Light does not move through the medium, nor does it require time, even an imperceptibly short time, to reach one extreme of the medium from the other. But we also know that no change can take place in an instant.

The presence of light simultaneously throughout the medium is due, Aristotle said, to the fact that light is not a species of motion: "For light is due to something existing, but is not a kind of movement" (446b28). He did not explain immediately what he meant by this, but a little later in the same chapter of *De Sensu*, he compared the actualization of light in the medium to freezing:

> And, in general, even in qualitative change, the case is not the same as in local movement. Local movements reasonably arrive at a midpoint first (and sound seems to be the movement of something moved locally). But for things subject to alteration, the case is different. For alteration may take place all at once, and not in half first, e.g., water may freeze simultaneously in every part. But, even at that, if what is warmed or frozen is great in size, each part is affected by the contiguous part, while the first part changes due to the agent of alteration itself, and it is not necessary that [all parts] together are altered all at once (446b28-447a7).

Like freezing, the actualization of light does not involve light's reaching a midpoint before it reaches its destination. Rather, it takes effect throughout the medium all at once. However, in this case, 'all at once' does not necessarily mean 'neither in an instant nor in a time period.' Freezing, as a case of alteration, is one form of movement and as such ought to take time. It is change in the category of quality and is restricted to change in the perceptible qualities of subjects which remain, themselves, substantially unchanged (*Physics* IV.14, 223a29-34; V.2, 226a23-29). Indeed, there is nothing in the passage that makes it certain that Aristotle was talking about light as change without the passage of time.[3] It would appear that we must look elsewhere for an explanation of the exemption of the motion of light from the requirements of time.

Philoponus took Aristotle's statement that light is not a species of motion as an indication that light is a true actuality, as opposed to a motion (*In de An.*, 296.25-297.10). This is why it requires no time to become actual throughout the medium. This distinction between actuality (*energeia*) and motion (*kinêsis*) originates in *Metaphysics* Θ.6, where Aristotle distinguished between actions which have a limit and those which lack one. If an action has a limit, this indicates that it tends toward a goal. Upon the attainment of that goal, the action ceases. Such an action is a movement, because it is not itself "that for the sake of which" the movement is undertaken. There are some actions whose end or goal is present in the action itself. These actions need not cease, since they do not tend toward a goal, like movement does.

It is a characteristic of such an action that we can use the present and perfect tenses of the verb describing this action, at the same time. For example, I can say at the same time, 'I see' and 'I have seen,' 'I think' and 'I have thought.' I would not say, 'I learn' and 'I have learned' at the same time, nor 'I am becoming healthy' and 'I am healthy' (1048b18-35). This passage is the primary reference for Aristotle's explanation of the distinction between motion and true actuality, which Philoponus invoked in reference to light. In finding the reason for light's propagation without time in the *kinêsis-energeia* distinction, Philoponus followed Alexander. Alexander did not refer to true actuality in precisely the terms Aristotle used in Metaphysics Θ.6. In his discussion of light in the *Mantissa*, pleasure is his reference for the relation of true actuality to time (*Mantissa*, 143.34). The discussion of pleasure is the context for Aristotle's reference to the *kinêsis-energeia* distinction in *Nichomachean Ethics* X.4. In this passage, Aristotle developed some of the implications of his characterization of actuality for the relation of pleasure to time. Aristotle said that pleasure is like seeing and different from motion, in that it is complete in any time whatsoever. One consequence of motion tending toward a goal is that motion is not complete in any part of the time it occupies. It is only complete when the end point has been reached and the movement has ceased. The many parts of a movement are incomplete, with respect to the whole. Consequently, since the form of a movement is determined by its starting point and ending point, any substretch of a movement differs in form from other substretches of the same movement. Pleasure, on the other hand, like true actuality, is complete in any time whatsoever, and any substretch of the total time occupied by the activity has the same form as the whole of which it is part. This is because an actuality requires nothing at a later time to perfect its form. Since the end is present in the activity itself, its form is complete in any substretch of the total period of true activity (1174a13-b9).

In commenting on Aristotle's use in *De Sensu* VI, 446b2 of the combination of tenses to describe the actualization of perception, Alexander pointed out the nature of the transition to true actuality. His remarks are similar to those of Aristotle in *Nichomachean Ethics* X.4:

> There are some things which do not proceed from non-being into being through generation; it is not possible to say of them that some part exists but nevertheless they are not perfected fully (*sumpeplêrôtai*), nor that they become wholes by further generation and by requiring some time for completion ... (*In de Sensu*, 125.12-15).

Alexander went on to say that the actualization of sense perception is precisely the sort of thing which is complete in any part of the total period of perceiving and which requires no time for actualization. He explained Aristotle's combination of tenses by reference to these characteristics, and he likened perception to pleasure in this respect (*In de Sensu* 125.23). Philoponus, in his commentary on *De Anima* II, referred to the *kinêsis-energeia* distinction in terms of its characterization in *Physics* III.1, 201b31:

> Actuality is, as he defined it in the *Physics*, the simultaneous advance from the state (*hê athroa probolê apo tês hexeôs*), while motion is incomplete actuality; for the road from the first potentiality to the state is motion (*In de An.*, 296.25-27).[4]

For the purposes of dealing with light, the implication that Philoponus drew from the *kinêsis-energeia* distinction was the same as Alexander's: the actualization of light throughout the medium in no time whatsoever (*achronôs*).

Were Alexander and Philoponus justified in ascribing the simultaneous actualization of light throughout the medium to its status as a true *energeia*? This question has two parts. Firstly, is it indeed the case that a true *energeia* is actualized all at once and without the passage of time? Secondly, would Aristotle have agreed with his commentators that light is a case of true actuality? Answering the first question is the subject of the remainder of this chapter. The second question is reserved for the next chapter.

II

Aristotle said nothing directly about the time of transition involved in the actualization of an *energeia*. To determine what is implied on that topic by his distinction between actuality and motion, we must define more accurately the meaning of the combination of present and perfect tenses used to characterize an actuality. The problem is understanding how the present tense entails the perfect for some verbs, and not for others. There are two positions on this issue, one represented by Gilbert Ryle and defended by Terry Penner, the other represented by J.L. Ackrill.[5] Since Ryle first drew attention to the importance of this issue in Aristotle, it is appropriate to begin with his interpretation.

Ryle found in Metaphysics Θ.6 his own distinction between achievement verbs and task verbs. An achievement verb, a 'got it' or 'missed it' word, involves the result of the action in its meaning, while for a task verb or 'try' verb, the result may be in doubt. Ryle's distinction is exemplified in the difference between treating or curing a patient, between running a race or winning it. Ryle thought 'seeing,' and other verbs of perception, signified achievements as opposed to tasks. He also thought that past philosophers, by ignoring this distinction, were led to hypothesize totally imaginary processes and mechanisms to explain perception:

> Special cognitive acts and operations have been postulated to answer to such verbs as 'see', 'hear', 'deduce', and 'recall' in the way in which familiar acts and operations do answer to such verbs as 'kick', 'run', 'look', 'listen', 'wrangle', and 'tell'; as if to describe a person as looking and seeing were like describing him as walking and humming instead of being like describing him as angling and catching, or searching and finding.[6]

Elaborating on this point, Ryle later tried to show why questions such as "How do we see trees?" or "What happens in us when we see trees?" predispose answers of the wrong sort, the kind which "assimilate seeing and hearing to having pangs and twinges."[7] The concepts of seeing and hearing should not be construed so as to answer questions appropriate to the sciences of physiology and psychology. Both these sciences study physical processes. 'Seeing,' however, signifies an achievement and therefore involves no notion of process. To explain what is involved in the concept of 'seeing,' Ryle called Aristotle to his aid:

> To begin with, seeing and hearing are not processes. Aristotle points out, quite correctly (Met. IX, vi, 7-10) that I can say 'I have seen it' as soon as I can say 'I see it'.[8]

Ryle took Aristotle to mean that 'seeing' is one of those verbs "part of the business of which is to declare a terminus." Such verbs are not used in the continuous present or past tenses:

> I can say that I am occupied in searching for a pencil or trying to solve an anagram, but not that I am occupied in finding the pencil

> or getting the solution of the anagram. In the same way I can be looking for or looking at something, but I cannot be seeing it.[9]

This tense usage for a verb like 'seeing' indicates that the verb signifies no process or experience that the seer undergoes. 'Seeing,' therefore, has no beginning, middle, or end. In fact, it is instantaneous. Ryle considered 'seeing' parallel to 'finding' or 'winning'; it is the "scoring of an investigational success." Consequently, there are two foci in Ryle's explanation of what Aristotle meant by the combination of tenses. The first is that 'seeing' entails 'having seen.' Seeing is not a stage on the way to something else, but involves a notion of the accomplishment of the action. The second focus is that seeing indicates an instantaneous action, the point of terminus to a process of looking. Like any end or beginning, "it cannot itself go on for a time, however short." Viewed in terms of the combination of present and perfect tenses, the two traits are closely related. Not only can I say 'I see it' and 'I have seen it' at the same moment, but also I say them about the same moment, the very one in which I see it. Ryle recognized that there are achievements that persist for a time, like keeping a secret. These verbs, while they involve a notion of success or accomplishment, do not declare termini, as Ryle believed 'seeing' does.

It is not clear whether Ryle believed the combination of tenses applied to both types of achievement verbs or only the instantaneous ones. In *Concept of Mind*, he mentioned Aristotle as one who recognized the difference between achievement verbs and process verbs.[10] This implies that the combination of tenses for Ryle might apply as well to achievements persisting through time. However, in *Dilemmas*, he considered *Metaphysics* Θ.6 only in relation to verbs that declare termini. Consequently, his justification of the combination of tenses in that context would not apply to achievements that continue through time. Even in *Concept of Mind*, it is the instantaneous achievement that most interested him. The problem is that in classifying 'seeing' as the termination of a process of 'looking for' or 'looking at,' he severely limited the meaning of a verb which, in its broader meaning, quite naturally would have fitted his characterization of achievements that continue through time.

J.L. Ackrill, in his investigation of *Metaphysics* Θ.6, agreed that for Aristotle, the combination of tenses meant that 'seeing' entails 'having seen.' But he was convinced, against Ryle, that Aristotle intended for seeing to persist through time. The first issue, that of entailment, Ackrill did not decide simply by an examination of the combination of tenses. He thought this combination, taken by itself, could imply simply the compatibility of 'seeing' and 'having seen' rather than entailment:

> Anyone, I suppose, will understand 'it is not true that at the same time one learns and has learnt' as meaning--not that 'learns' does not entail 'has learnt' but--that 'learns' and 'has learnt' are incompatible. Surely therefore the contrasted formulae like 'sees and has seen' should mean simply that these pairs are compatible and can be true at the same time.[11]

In deciding in favor of entailment, he relied on other evidence, in particular *Metaphysics* Θ.6, 1048b18-23, the passage preceding Aristotle's presentation of the combination of tenses. In this passage, Aristotle drew the distinction between actions which have a limit and those which lack one. Ackrill explained that the limit "points both ways":

> If you have already built a house you cannot still be building it; but also, if you are still building it you cannot yet have built it. There is a limit which both cannot be passed and must be reached. So, correspondingly, an action which altogether lacks a limit should be one not only continuable indefinitely but also fully performable in any period of time however short.[12]

Seeing is complete throughout the time period of the activity. When this interpretation is applied to the combination of tenses, 'seeing' entails 'having seen' because of the preceding period of 'seeing.' If 'seeing' did not entail 'having seen,' then completeness would be lacking at the earlier stage. As Ackrill expressed it in general terms:

> If it could be true to say that a man *X*es while denying that he has *X*ed, the activity *X*ing would not 'lack a limit'; the man would still be on his way to a limit, viz. the point at which it would be true to say that he has *X*ed.[13]

It must be clear by now that this notion of entailment is very different from Ryle's. Taking *X* as 'see,' for Ackrill '*X*es' entails 'had *X*ed,' means '*X*es now' and '*X*es at a previous time.' On the other hand, for Ryle '*X*es' entails 'has *X*ed,' means '*X*es now' and 'has *X*ed now.' Ackrill himself articulated the working difference between the two: "While Ryle's account of the present-perfect connection involves that an *energeia* cannot go on through time, this one implies that it must."[14] Underlining his position, Ackrill referred to *Physics* VI.5 and 6, where Aristotle showed that there is no first subperiod in a period of change, and no first moment of change.

In this passage, Aristotle argued that there can be no first time in which something is changing. The first time in which something is changing cannot be an instant, since there can be no motion in an instant, as we have already seen. Furthermore, if the first time of movement were an instant, then there would be two successive instants, the limit of the preceding period of rest and the first moment of motion. But instants cannot be successive. Consequently, if the first time in which something is changing is an instant, the object is both changing and not changing at the limit between the period of rest and movement. But this is, of course, impossible. The first time in which something is changing cannot, therefore, be an instant. But neither can it be a period. If the first time in which an object KL is moving is the period XP, it is still possible to imagine another object, moving with equal speed, which covers half the same distance as KL in half of the period XP. Then, XP is not the first time of KL's movement at all (*Physics* VI.6, 236b34-37a4, 237a17-25). In general, no matter what segment XP of a whole motion is taken as the first time something is moving, it is always possible to choose within that segment another smaller segment XM in which the object is moving before (*Physics* VI.5, 236a20-28). Ackrill thought that these same arguments could be applied to true actuality: there is no first subperiod or first moment in a period of such actuality. He compared motion and actuality on the basis of whether or not it is possible, while the activity is taking place, to use the perfect tense to describe the action taking place during a previous subperiod of the whole period of the action.

Ackrill's interpretation of the combination of tenses relied on a previous period of 'Xing' for the meaning of 'has Xed.' He concluded from this that there can be no first moment of a true *energeia*. In Ryle's interpretation, 'has Xed' applies to the very same instant as 'Xes.' Based on his interpretation of the way the present tense entails the perfect, there could be a first moment of a period of true actuality. While Ryle thought the combination of tenses describes only instantaneous actions, it seems clear that Aristotle thought his *energeiai* describe actions that continue over time. It would appear that Ryle's opinion should be rejected on this basis alone, in favor of Ackrill's. However, Ackrill's interpretation has a crucial flaw in its failure to consider the implication of *Nichomachean Ethics* X.4 that an *energeia* can take place in an instant. If this is so, it means that the 'has *X*ed' used to describe an actuality cannot refer to a previous period of '*X*es,' as Ackrill thought.

The suggestion that true actuality can occur in an instant arises from Aristotle's comparison of the forms of motion and pleasure:

> It seems that [motion] is not perfect in every time, but the many parts [of motion] are imperfect and differ in form, if it is true that the whence and whither constitute its form. But the form of pleasure is perfect in any time whatsoever. It is evident then that motion and pleasure are different from one another, and that pleasure is a whole and something perfect. This would appear to be the case also from the fact that motion is not possible except in time, but not so with pleasure; for what is in a moment is a whole (1174b4-9).

On the basis of this statement, Penner argued against Ackrill's idea that there can be no first instant in true actuality. Starting from Aristotle's statement, *to pothen poi eidopoion* (the whence and whither constitute the form), he sought to establish the conditions of identity for a motion, as opposed to an actuality.[15] From Aristotle's presentation, in *Physics* V.4, 227b23-26, of the criteria which must be satisfied for a movement to be one in number, Penner concluded that different movements are the same in form, if they start at the same place and end at the same place. Before it can be said that two movements are numerically the same, it must be determined that they also start at the same time and end at the same time. All comparisons of movements are based on these conditions of time and place. Because of the infinite divisibility of time and place, it is always possible to define a movement shorter than any given movement one might choose. On this principle depends Aristotle's argument that there is no first subperiod of change. However, the same argument cannot apply to an *energeia*, since it is not dependent on the same criteria of identity. Since an actuality contains its end in itself, no notion of departure from one point and arrival at another is involved in the definition of its form. Consequently, it is inappropriate to apply to an actuality the argument that there can be no first subperiod or first moment of its period of duration. Since this parallel cannot be established, it is quite possible that there is a first moment of actuality. Citing *Physics* VI.5, 236a14-27, where Aristotle established the incompatibility of a last moment of rest and a first moment of movement,

Penner concluded that it is purely arbitrary whether we say there is a first moment of actuality or a last moment of its unactualized state.[16]

This point is important for understanding the commentators' use of the concept of actuality in relation to light, because it establishes that there can be a first moment of the presence of light in the medium, which is preceded by no period of change. To understand why there must be a first moment of the presence of light and not a last moment of the unactualized transparent, we must determine precisely how motion and actuality are comparable. When Aristotle said that true actuality requires nothing at a later time to perfect its form (*Nichomachean Ethics* X.4, 1174a15-16), he stated in reverse Ackrill's explanation of an *energeia*: that it is already complete in any given subperiod of the larger whole period of the activity. Ackrill's description of what is an *energeia* is basically correct[17] and harmonizes with Penner's insight on the problem except for certain corollaries--for example, whether or not an *energeia* may be in an instant. Ackrill's mistake, on this point, is due to his reliance on a previous period of '*X*es' for the meaning of 'has *X*ed.' However, the perfect tense in Greek, as in English, most properly refers to action completed in the past whose effect continues into the present. If the perfect tense involves reference to past action, how can the perfect tense of a verb apply at all to the first moment of the activity described by that verb? We can answer this question by noticing the way Aristotle used the perfect tense for both movements and actualities.

In the analysis of motion, the perfect tense signifies the achievement of the end; the first moment in which something 'has changed' is the first moment in which the change has been completed (*Physics* VI.5, 236a7-8). This is why Aristotle argued, in *Physics* VI.5, 235b19-30, that in a change from A to B, something has not changed until it is in B. If it has changed on the way to B, it 'has changed' and 'is changing' at the same time. We can agree with Aristotle that this is impossible only because we understand the perfect tense necessarily involves the notion of its goal. For example, if a man is building a house, but has only finished the west wing, we cannot say he has built until he has finished the house. We may say he has built the west wing, but then the identification of the motion has been altered. A verb signifying a motion always involves the idea of its direct object as the goal; the perfect tense signifies the achievement of the goal implied by the direct object and the consequent cessation of the activity.

Turning to the analysis of activities in *Metaphysics* Θ.6 we see that the perfect tense has exactly the same meaning as it does in *Physics* VI.5. It signifies the achievement of the goal. That the goal of an activity is present in the very exercise of the activity has as a consequence that the activity need not cease (1048b26). This characterization of activity yields, upon analysis, Ackrill's understanding of a period of activity: " ... that there is no upper or lower limit to the time it may occupy, and that it is somehow equally and fully present throughout any such period."[18] However, it is important to retain Aristotle's formulation, because it is a key to the technical way in which he understood the perfect tense in his comparison of motion and activity. Motion ceases because there is a *telos* toward which it is tending. The ability of an *energeia* to continue indefinitely is an extremely important point because it is a practical result of the primary distinction between the two types of activity. Since it is the perfect tense which signifies the achievement of the goal, and the goal is ever present in an activity, it must be possible to use the perfect tense to refer to any period or moment within the activity. In making his distinction, Aristotle acted on the recognition that some verbs in combination with their direct objects can take the present and perfect tense at the same time, and that others cannot. This does not mean, however, that he began from any irreducible datum such as the fact that when someone has been looking for something and finds it, he says 'I've found it,' when actually *he is just finding it*.[19] We do not need such an inscrutable starting point to understand what Aristotle saw in the difference between verbs of motion and activity. It is more likely that, for Aristotle, the fact that the present and perfect tenses can be used at the same time signified primarily that the activity can continue indefinitely. From this point of view, what is remarkable is the fact that the present can be used when the perfect applies, rather than *vice versa*. This is remarkable because the perfect most properly implies completion. It is in this sense of completion, derived from the analysis of motion but applied as well to activity, that the perfect tense applies to the first moment of an activity. In considering verbs whose present and perfect tenses may be used at the same time, Aristotle understood the perfect tense to refer to the achievement of the goal. Apart from this, however, the combination of tenses conveyed to Aristotle primarily that the action could continue indefinitely.

Up to now, we have compared actuality to motion. However, as the preceding discussion indicates, a more natural counterpart to actuality is the period that commences with the completion of motion. The completion of motion shares with actuality the presence of the end for which the action was initiated. Bearing this in mind, let us reexamine Aristotle's argument in *Physics* VIII.8, 263b9-26, against change in a moment. In this passage,

Aristotle dealt with the problem of the limit between the period in which an object is non-white and the ensuing period in which it is white. The limit, insofar as it belongs to both periods, must be both non-white and white. Aristotle solved this problem by assigning the limit to the later period, saying that throughout the earlier period of being non-white, the object is in the process of changing to white and the limit marks the achievement of the goal of that change and the commencement of the period of being white. Transposing the change from state to actuality onto this model for motion, we see that actuality fits perfectly the commencement of the period of being white. The first moment of actuality matches the moment of completion of change. The end is present in this first moment of the actuality, just as the goal has been reached at the moment which marks the completion of the motion. The only difference is that the period of change supposed by Aristotle to precede the achievement of the goal has been eliminated.

If light is an *energeia* of the kind under discussion, then let us consider the first moment of the presence of light in the medium. Light is the actuality of the transparent, which before the presence of a luminous object in the medium, exists only potentially. If there is a first moment of light, there is no last moment of the unactualized transparent, and the change from potential transparent to light takes place in no time whatsoever.

III

If the transition from darkness to light takes place without the passage of time (*achronôs*), how is this different from change in an instant (*tôi nun*)? For Aristotle, instants serve as markers in time and possess no magnitude. They are not indivisible times. Neither Alexander nor Philoponus discussed the meaning of *achronôs* in relation to vision or light. However, the author of the commentary on *Metaphysics* Z.8 which is ascribed to Alexander equated *achronôs* and *en atomôi nun*, in a discussion about motion.[20] Before examining his remarks, let us return to the comparison between light and freezing in *De Sensu* VI, 446b28-447a7. It was pointed out that this comparison cannot be relevant to the liberation of light from any time reference, because freezing, as a form of alteration, must take time. But in what sense may alteration take place all at once, if it must take time? To investigate this question will reveal some of the limitations of Aristotle's analysis of continuous change and thus make more readily understandable the resolution of the difference between *achronôs* and *tôi nun* offered by the author of the commentary on the *Metaphysics*.

In the passage from *De Sensu* VI, Aristotle likened the advent of light in the medium to freezing, an alteration which is accomplished all at once (*athroon*). It is worth noting that *athroos* evolved as a descriptive word for a collectivity and had, originally, no reference to time. The basic meaning, 'as a mass' can be found with remarkable regularity in Aristotle's other uses of *athroos* and its variants. The word occurs most frequently in the biological and natural philosophical works in reference to collectivities, such as masses of air or water.[21] However, even when not used in direct reference to collectivities, the word retains some notion of 'as a mass.' One example of Aristotle's usage of *athroos*, which is particularly illuminating in relation to the problem of change without time, occurs in *Politics* XI, 1315a13, where Aristotle discussed depriving a tyrant of his power: "And if anyone is to be deprived of his power, let it be removed gradually (*ek prosagôgês*) and not taken from him all at once (*athroon*)." While at first glance, *ek prosagôgês* and *athroon* may appear to comprise a contrast as to time, closer consideration shows that it is the power which is removed gradually or all at once. In the case of *athroon*, the power is taken away 'as a mass.' The meaning of *prosagôgê* bears out this interpretation, insofar as it refers to introduction or addition in terms of a thing, not in terms of time. Even in this contrast between 'all at once' and 'gradually,' the reference to the mass, rather than to time, is apparent.

Having gotten this far, it is important not to leave out entirely the issue of time. Even if the tyrant's power is removed, that it is removed gradually implies a certain relation to time. When the removal is made all at once, all the power is taken away at some one time (*pote*). When the removal happens gradually, power is taken away bit by bit at a series of different times (*pote*), or it may possibly be reduced continuously, the process thereby corresponding to a certain period of time (*tis chronos*). That the removal of power at some time might also correspond to a period of time, however short, is not a point which is relevant to the contrast Aristotle made in this passage. The relation to time implied by *athroon* is completely informal and could have been expressed by the word *pote*, had Aristotle wished to clarify it.

Thus, it would appear that *athroos* implies a non-technical designation of time: a specific time, but one indefinite as to duration. To determine if it has a more specific meaning in relation to freezing, let us consider its role in a passage more central to Aristotle's general analysis of motion, in particular *Physics* VIII.3, 253b13-254a1, a passage which treats alteration that happens all at once.[22] The problem of *athroa alloiôsis* arose in the context of refuting the idea that all things are always in motion (253b6). Aristotle argued against this idea by showing that change is not, in every instance, continuous. Growth and diminution cannot be continuous because there must be some time between the two processes not typified by either motion. One process cannot, without ceasing, become the other (b13). A separate point raised against the same idea, that all things are always in motion, is the comparison of the process of diminution to hauling a ship. He likened the combined efforts of hauling a ship to the effect of water dripping on rock and gradually wearing it away. If a certain amount of rock is worn away in a certain time, it need not follow that half the amount is removed in half the time. It may not be the case that the rock gets smaller with each drop, but many drops may be required to produce any effect and then the effect may be grand in scale:

> Then, what is removed is divisible into a number of parts, but none of them is moved separately, but all together. It is evident then that it is not necessary, because the wearing away is infinitely divisible, that some part is always being removed, but rather the whole is removed at some time (253b19-23).

It is important to notice what Aristotle meant by *holon pote apienai*, because when he went on to speak in similar terms about alteration, it was this 'removal at one time' which was comparable to *alloiôsis athroa*:

> Similarly, too, for any alteration whatsoever, it is not the case if what is altered is infinitely divisible, for this reason also the alteration [is infinitely divisible] but [the alteration] often occurs all at once, just like freezing (253b23-26).

Any subject of alteration is infinitely divisible, as is any subject of diminution. But it does not follow from this that either alteration or diminution must be successive (and consequently continuous with respect to time), because diminution might happen *holon pote apienai* or alteration might happen *athroa*.

It is important to notice that Aristotle contrasted the continuity or suddenness of change in terms of the continuity of the subject of change.

While this is natural in the case of the rock's erosion, it does not fit alteration so well. That freezing takes place all at once means not only that the subject is not affected successively, but also that cooling is not continuous in its effect of producing solidity. This is easy to accept, in the case of freezing, because solidity does seem to take effect all at once, without an intervening period of progressive thickening. But are there alterations that take effect gradually throughout the subject as a whole? Certainly, one would expect that in a change from black to white, for instance, the gradations in shade between the contraries would make their appearance in time. However, Aristotle's remarks on the subject would indicate that even these gradations take effect only part by part.

In *Physics* VI.9, 240a19-29, Aristotle argued that when something is in process of changing from non-white to white, some of its parts must be white while others remain non-white, because while it is changing, it cannot be neither non-white nor white. This is the sense in which change takes place part by part. More and more of the subject becomes white rather than non-white, until all of the subject has passed into the latter state and the change is complete. This argument has the same basis as his discussion, in *Physics* VIII.8, 263b9-23, about the character of the instant separating the periods of non-white and white. When the object is changing, it cannot already be in that into which it is changing. It cannot be both non-white and white, but also it cannot be neither. In *Physics* VIII.8, he resolved the problem of the limit between the periods of the object being non-white or white by assigning the instant to the later period. In *Physics* VI.9, he considered the object itself, rather than the moment when the change is complete. He concluded that we judge to which state the object belongs by seeing whether most of its parts are non-white or white. Aristotle generalized this point about change part by part and applied it to any kind of change at *Physics* VI.4, 234b10-20. But he made an important qualification, applying the doctrine to stages of change within a larger change. Insofar as any movement AD is composed of a series of smaller movements AB, BC, and CD, at any given stage of the movement BC, part of the mobile must be in B and part in C, but it cannot be wholly in both B and C. But no part of the object need be in A or B, even though these are the extremes of the change in question. Aristotle himself showed how to apply the idea to the change from white to black, considering the change from white to gray only.

This qualification helps us to understand the doctrine better, because it deals with alteration as something other than a change from non-white to white. Aristotle's explanations of qualitative change are often confusing, because his arguments treat changes between contraries as if they were changes between contradictories. However, his adherence to the not-A to A model is more appropriate than it may appear to us at first glance. Any change in shade along a scale between contraries, like black and white, may be construed as a change from not-A to A, because of the discontinuous nature of all series of colors.

Perceptible qualities (the class in which alteration takes place) are in all cases not subject to infinite divisibility. Aristotle said, in *De Sensu* VI, 445b21-23, that the kinds of color, sound, and other sensible objects are limited. Between the extremes of black and white, there is a limited number of shades. The shades, or *eidê*, of color form a finite series. Though the shades are not continuous, yet "continuity always inheres in them" (445b30). Illustrating this principle, Aristotle referred to the imperceptibility of the quarter-tone on the musical scale, in spite of its contribution to the whole interval (446a1-4). Based on this passing remark, Sorabji has made a reasonable conjecture about the way continuity inheres in the discontinuous series of perceptible qualities:

> What Aristotle seems to have in mind is that a change to the next discriminable pitch, in the discontinuous series of discriminable pitches, may be produced by a continuous movement of a stopper along a vibrating string. Or in the case of colour, a change to the next discriminable shade, in the discontinuous series of discriminable shades, may be produced by a continuous change in the proportions of earth, air, fire and water in a body .... A body is changing to the next discriminable shade all the time that the continuous change in its elemental ingredients is going on, which will eventually lead to its displaying that next discriminable shade.[23]

In this way, a subject may be changing to the next shade, even though there is no sign of this until the change is complete. This hardly makes alteration change along a discontinuous series, but rather continuous change that appears discontinuous. However, Aristotle was interested in qualitative change insofar as it is perceptible. To this extent, the material basis that produces the series of colors or sounds is incidental. Insofar as the senses experience color or sound, these series are truly discontinuous. Only visible objects or producers of sound experience the material changes that produce the alteration in their perceptible qualities.

The perceptible qualities of objects, therefore, change discontinuously. This means that any change in shade happens all at once for any given part of the changing object. It would be appropriate to consider any change between shades as a change from not-A to A, even if the extremes involved are, for instance, white and gray. We may understand any discontinuous change as change from one atom of color to another. Then, how does such qualitative change take time? We could fall back on Aristotle's doctrine that alteration takes place part by part to insure that it takes time. However, that may not be necessary, if we remember Aristotle's statement in *Physics* VIII.8, 263b21, that while the object is yet non-white, it is changing to white. This statement makes sense if we remember the way in which continuity inheres in the discontinuous series of perceptible qualities. The object itself is changing, before that change becomes manifest to the senses.

Aristotle's reference to freezing in *De Sensu* VI, 446b28-447a7, provides two sorts of alteration: one which happens all at once, and the other part by part. Even gradual alteration (white to gray to black) may take place part by part. Theoretically, there could be gradual change throughout a whole as well, though Aristotle never mentioned this possibility. The point is that whether alteration is made gradual by succession through the subject or by progress through a series between contrary qualities, the process is merely an itemized and multiplied version of *athroa alloiôsis*, the paradigm case of which is freezing. Freezing is the production of solidity due to the accumulation of the effect of cooling. In this way, it is not so different from the accumulated change in the material conditions on the surface of an object that results in a change of color. Any alteration, even a stage on the way to a more extreme change, can be represented by *athroa alloiôsis*.

It seems that rather than making the actualization of light more readily understandable in comparison to alteration, our analysis of the comparison to freezing has rather reduced alteration to a series of discontinuous changes. The problem lies in Aristotle's inability to dispel the change from not-A to A that 'ghosts' the first moment of the completion of change. It is this 'ghost-change' that, in the case of true *energeia*, led both Philoponus and the author of the *Metaphysics* commentary to say that the transition from potentiality to actuality takes place in no time whatsoever. It is interesting to see how the author of the *Metaphysics* commentary treated the problem in relation to movement, however.

In his commentary on *Metaphysics* Z.8, 1033a24, pseudo-Alexander equated *achronôs* and *en atomôi nun* (*In Metaph.*, 495.23) while discussing how the form comes to be present in an object at the conclusion of change. He said that the form is not generated a little at a time in an object, but that it "supervenes in it without the passage of time and in an indivisible instant." Pseudo-Alexander made this point for cases of ordinary movement, like housebuilding. He said that a house has not come into being until the final brick is in place. At that point, the form supervenes and completes the house, in an instant (*In Metaph.*, 486.17-22). Pseudo-Alexander was referring to Aristotle's first moment of 'having changed.' This moment marks the completion of the change and is also the first moment of the new condition. Accordingly, the advent of the form in no time whatsoever occurs at the very instant which, Aristotle specified, belongs to the period of actualization that succeeds the preceding period of 'changing.' This coincidence of transition and completion in the same instant need not be paradoxical, if we understand properly in what sense something can 'take place' in an instant.

Instants possess no magnitude. This is the most important point for evaluating the two arguments against motion in an instant, presented at the beginning of this chapter (*Physics* VI.3, 234a24-31; VIII.8, 263b9-26). The upshot of the first argument is that, for an instant to be indivisible, it must lack magnitude, serving only as a marker in time. This is the function of the instant in the second of Aristotle's arguments against motion in an instant. The main point of this argument is that there is no time in which an object in change is neither what it was nor what it shall be; there is no intermediate state. However, the implication for the limit of time between the two states is that it can have no identity independent of the period of time to which it belongs. An instant is not a species of time, just like a period, but rather is parasitic on time for its reference in place or condition. If instants are not part of time and serve only as markers, what is the significance of discussing to which of the periods the limiting instant belongs? It reflects the nature of change. There can be no last moment of change. There can, however, be a first moment of the completion of change. The reason is that change and remaining unchanged cannot each be identified in the same way at an instant. Even if instants serve only as markers, still they mark something. But in change, identity is defined by point of departure and destination. Consequently, an object can only be defined in terms of progressively smaller locations marked by limits progressively nearer to one another. One could never identify what corresponds to an instant of change except in this way. And this manner of identification would not constitute an 'instant' of change at all. In contrast, the completion of change is characterized by a form the object possesses throughout time. Since this form is constant throughout time, it is possible to define it completely, at any given instant, including the initial limit of the period.

As a result, when we say that a true actuality, following the example of pleasure in *Nicomachean Ethics* X.4, can take place in an instant, this does not mean that the action transpires during the course of an instant, but simply that the nature of the action can be fully defined at an instant. When Alexander said the form, in motion, comes to be without the passage of time and in an indivisible instant, he said this is because the form is indivisible. It cannot make its appearance gradually. There is a transition, a process of change leading up to the advent of the indivisible form. The form only completes this change by supervening in the finished product, and this supervention is not a form of change at all. The time when this indivisible form is first present in a thing can be an instant, because the completed state of the object is the kind of thing which is constant through time. Its identity is not dependent on any time-related references, like points of arrival and departure, for its characterization. Consequently, it can be present in an instant. This is what it means to say the form supervenes in an instant. It would be appropriate, in fact, to translate *achronôs* as 'without time,' since the supervention of the form at the end of a process of change is not a species of movement, and hence need not have reference to time at all. This is consistent with Alexander's statements that light does not need time for its actualization (*In de Sensu*, 132.1). However, Philoponus (and also Alexander, *In de Sensu*, 133.3-6) used *achronôs* in reference to the fact that light can travel enormous distances all at once. Consequently, it is unmistakable that both Alexander and Philoponus used *achronôs* to mean 'without the passage of time' or 'in no time whatsoever.' It would appear that the explanation that an *energeia* requires no time for its actualization lies behind the fact that light can cover any distance in no time whatsoever. Indeed, it is only by this explanation that Philoponus escapes the paradox that light comes to be present throughout the medium neither in an instant nor a time period.

IV

We have seen that *athroos*, in the case of alteration, means that the perceptible change takes place in no time, as a change from not-A to A. During the period that the object is in the first state, it is changing imperceptibly in such a way as to produce the final dramatic change. The actualization of light in the medium differs from freezing only in lacking this foregoing period of 'changing.' Freezing, the paradigm of alteration, fits perfectly the general case of change in *Physics* VIII.8, 263b9-26, elaborated at the beginning of this chapter. It also points up the limitations of Aristotle's analysis of motion for dealing with sudden or instantaneous change. He never reconciled this sort of change to his model of continuous change. For cases of change all at once, he merely shunted the actual moment of change into a no-man's land that corresponds to no time reference at all. For changes that are genuinely continuous, like locomotion, he can avoid this problem. But there are changes which cannot be conceived as continuous (freezing) and others which cannot take time (illumination). *Athroa alloiôsis* does not escape the universal maxim that motion takes time, but Aristotle must supply a period of imperceptible change to make it so. The change from darkness to light takes place outside of time, but only because of its association with the class of actuality that contains its end in itself. And this fact only comes to light in the hands of the commentators.

Notes

1. See, for instance, his commentary of *Physics* 253b26 (vol.15 and 16 in *Comm. in Arist. Gr.*, ed. H. Vitelli) 883.33-884.2.1.

2. We will return to this general case of movement more than once in this chapter. For an important treatment of the significance of this passage in Aristotle's analysis of motion, see G.E.L. Owen, "*Tithenai ta phainomena*," *Aristotle*, ed. J.M.E. Moravcsik (Garden City, New York, 1967), 183-87; reprinted in *Articles on Aristotle*, ed. by J. Barnes, M. Schofield, and R. Sorabji (London: Duckworth, 1975), vol.1 (Science), and *Logic, Science, and Dialectic: Collected Papers in Greek Philosophy*, by G.E.L. Owen, ed. by Martha Nussbaum (Ithaca, N.Y.: Cornell University Press, 1986).

3. To say that light, being an alteration, does not reach a midpoint cannot by itself establish that the change of which light is the result takes place without the passage of time. There is a difference between asking if light reaches a midpoint in the medium before reaching the eye, and asking if an interval of time exists between introduction of a light source in the medium and vision. While the former question necessarily implicates time, it applies less generally than the latter question. There may be no contrast in this passage between motion which takes time and alteration which does not, but rather between successive change and change 'as a whole.' This is, in fact, all the meaning that Alexander drew out of it (*In De Sensu* (vol. 3, part 1 in *Comm. in Arist. Gr.*, ed. P. Wendland) 133.1-6).

4. Philoponus always invoked the distinction in terms of the incompleteness of the actuality of motion. He emphasized the nature of the potentiality and actuality involved in motion and true actuality respectively. Because of this emphasis, the *kinêsis-energeia* distinction became an important element in his doctrine of first and second potentialities and actualities. This doctrine is treated in Chapter Four.

5. Gilbert Ryle, *The Concept of Mind*, (Harmondsworth, Middlesex: Peregrine, 1963), 143-47, and *Dilemmas* (Cambridge: Cambridge University Press, 1960), 143-47; Terry Penner, "Verbs and the Identity of Actions--a Philosophical Exercise in the Interpretation of Aristotle," *Ryle*, ed. by O.P. Wood and G. Pitcher (Garden City, N.Y.: Anchor Books), 393-460; J.L.

Ackrill, "Aristotle's Distinction Between *Energeia* and *Kinêsis*," *New Essays on Plato and Aristotle*, ed. by Renford Bambrough (London: Routledge and Kegan Paul, 1965), 121-141.

6. Ryle, *Concept of Mind*, 145.

7. Ryle, *Dilemmas*, 107.

8. Ryle, *Dilemmas*, 102.

9. Ryle, *Dilemmas*, 103.

10. Ryle, *Concept of Mind*, 143.

11. Ackrill, "*Energeia* and *Kinêsis*," 123.

12. Ackrill, "*Energeia* and *Kinêsis*," 124.

13. Ackrill, "*Energeia* and *Kinêsis*," 124.

14. Ackrill, "*Energeia* and *Kinêsis*," 127.

15. Penner, "Identity of Actions," 403-4, 411-24.

16. Penner, "Identity of Actions," 410.

17. Ackrill's explanation is reinforced by Fred D. Miller, Jr.'s investigation of the meaning of oneness in Aristotle's philosophy ("Did Aristotle Have a Concept of Identity?" *Philosophical Review* 82 (October, 1973), 483-90. He suggested that "oneness over time amounts to unitariness over time in the case of processes, and to identity over time in the case of activities and substances" (489). He believed this difference is "at the root of the division of events into processes and activities." What is most interesting about his investigation is that, while affirming Penner's position on the identity criteria of motion, Miller arrived at identity criteria for actuality which fit precisely Ackrill's explanation.

18. Ackrill, "*Energeia* and *Kinêsis*," 128.

19. Ryle's insight on the verb 'seeing' was elaborated by Zeno Vendler ("Verbs and Time," *Philosophical Review* 66 (April, 1957), 143-60) in the following way:

> ... in cases of pure achievement terms the present tense is almost exclusively used as historic present or as indicating immediate future. 'Now he finds the treasure (or wins the race, and so on)' is not used to report the actual finding or winning, while the seemingly paradoxical 'Now he has found it' or 'At this moment he has won the race' is (147).

For Ryle, seeing is like finding or winning, in Vendler's explanation. Better understood as 'spotting,' it is the culmination of a previous process, namely 'looking for.'

20. Alexander, *In Metaphysica* (vol. 1 in *Comm. in Arist. Gr.*, ed. by M. Hayduck), 495.23.

21. *De Anima* II.8, 420a25, *Meteor.* II.2, 354b3-15; in *Meteor.* IV.9, 387b27, the presence of moisture in wood is described as *athroon* and *sunekês.* In *Gen. An.* V.3, 783b14-17, *athroos* is used in reference to molting and balding. See also *Gen. An.* III.5, 756a11-14, where *athroon* is used in relation to *kata ton auton chronon.*

22. Owen thought that, in this passage, Aristotle "agrees that some changes take no time at all" ("*Tithenai ta phainomena,*" 185). For criticism of Owen's interpretation, see Appendix II.

23. Richard Sorabji, "Aristotle on the Instant of Change," *Proceedings of the Aristotelian Society* 50 (1976): 80; reprinted in *Articles on Aristotle,* vol.3 (Metaphysics), 159-77.

Chapter Four

The Propagation of Light, Impetus Theory, and Aristotle's Physics *III.3*

I

Aristotle never stated directly that light is the sort of actuality that contains its end in itself. There is every reason to believe, however, that he would have agreed with Alexander and Philoponus when they stated this fact directly. In the first place, sense perception--especially visual perception--was one of Aristotle's favorite examples of true actuality. And there is no doubt that light follows sense perception in being modelled on the doctrine of capacities found in *De Anima* II.5. In *De Sensu* II, 438a14-15, Aristotle even asserted that vision occurs insofar as the water in the eye is transparent. Furthermore, in *De Sensu* VI, he invoked the combination of tenses, as applied to sense perception, in comparing the progress of sound and smell through the medium to the actualization of light:

> Even if one hears and has heard all at once and, in general, perceives and has perceived, and there is no generation of them, but they exist without being generated, nevertheless it is no less the case that sound, for instance, when its source has already sounded, has not yet arrived at the sense of hearing (446b2-6).

Even though the senses of hearing and smell are activated all at once, sound or smell takes time to reach the senses. Aristotle made this point precisely to contrast sound and smell to light (446b10). Light is generated throughout the medium with the same simultaneity implied by the combination of tenses that can be applied to vision.

Besides examining Aristotle's specific remarks about light, let us consider if light is an actuality because of its reliance on the doctrine of capacities in *De Anima* II.5. In this passage (417a21-b27), Aristotle divided potentialities into two groups: the power to acquire a capacity, and the power to exercise a capacity already acquired. The former is exemplified by the ability to learn, and the latter, by the ability to exercise knowledge which one already possesses (a21-30). Aristotle said that no change takes place when a capacity is exercised, because the exercise of a capacity is a realization of an ability already possessed. Philoponus thought that all cases of the actualization of a capacity already possessed are also cases of true *energeia*. On this basis, light, as the actualization of transparency, is a true *energeia*. In effect, Philoponus allied the distinction of *Metaphysics* Θ.6 with the doctrine of capacities in *De Anima* II.5. In his interpretation of *De Anima* II.5 and related passages, Philoponus originated the causal theory common to his explanations for the propagation of light and projectile motion. It has been supposed that the doctrine of imparted power was carried over into natural philosophy from Neoplatonic metaphysics.[1] However, Philoponus' theory of imparted power derives from his analysis of the way in which the activity of agent and patient are one, for either a movement or actuality. As the following analysis of *De Anima* II.5 will demonstrate, this is one of the central issues raised by the relation between *De Anima* II.5 and *Metaphysics* Θ.6.

In *De Anima* II.5, Aristotle considered two types of potentiality that might be possessed by a subject: the power to acquire a capacity and the power to exercise a capacity already acquired. Terms such as *to paschein* and *alloiôsis* imply some type of motion. It is not appropriate to apply these terms to exercising a trait unless we distinguish two senses of 'being acted upon':

> Nor is 'being acted upon' used in one way, but on the one hand, it is a kind of destruction by the agency of a contrary, and on the other hand, it is rather a preservation by what is actual of what is potential and similar to it as potentiality is related to actuality (417b1-5).

It is more appropriate to apply *to paschein* and *alloiôsis* to the acquisition of a capacity, because they involve a transition between contraries. However, Aristotle qualified the application of these terms to the potentiality for acquiring a capacity:

> That which, from being potential, learns and acquires knowledge by the agency of what is actual and capable of teaching either ought not to be called 'being acted upon,' as was said, or else there are two kinds of alteration, one a change to conditions of privation, the other a change to states and nature (417b12-16).

Even though the acquisition of a capacity is a change between contraries, it is less appropriate to call it an alteration than some other such changes. Learning is a change from ignorance to knowledge of a subject. It is the acquisition of a positive state, made possible by the potentiality present in the subject. It differs in this way from a mere change between contrary qualities, like white and black, or hot and cold. One of these contrary qualities is destroyed by the advent of its opposite, but loss of ignorance on the part of a subject is not so much a destruction as "an enhancement of its natural properties."[2]

Aristotle thought that the actualization of neither kind of potentiality could be described as alteration as readily as some other sorts of change. The main point at issue here, however, concerns the power to exercise a capacity. By qualifying the use of words implying change to describe exercising a capacity, did Aristotle mean that the change typified by the transition from the possession of knowledge to its actual exercise culminates in a true actuality as opposed to a movement? Certain indications make us think so. In the first place, Aristotle referred to the doctrine of *Metaphysics* Θ.6 just before his discussion of the two principal types of potentiality:

> First, then, let us speak as if 'being acted upon' and 'being moved' and 'actualizing' are the same, for movement is a kind of actuality, though imperfect, just as was said elsewhere (417a15-17).

In the second place, the examples in *De Anima* II.5 of the exercise of a capacity already possessed, 'thinking' and 'perceiving,' happen also to be examples of true actuality in *Metaphysics* Θ.6. Unfortunately, Aristotle's reference to the distinction between motion and actuality is not taken up in the discussion of the two types of potentiality which follows it. And even though *De Anima* II.5 and *Metaphysics* Θ.6 have some examples in common, not all cases of the exercise of a capacity qualify as true actualities. In *De Anima* II.5, 417b8-9, Aristotle presented two examples of a capacity already possessed, but one of them we know is a movement and the other an actuality. In the passage leading up to b8-9, Aristotle said that the exercise of knowledge, as the preservation of what is potential by the agency of what is actual and similar to it, is not alteration: "For it is a development (*epidosis*) into itself and into actuality (*entelecheia*)" (417b6). He added, "Thus, it is not right to say the thinker is altered, when he thinks, just as neither [is it right to say] the housebuilder [is altered] when be builds" (b8-9). Thinking is a prime example of a capacity already possessed, and it is also one of Aristotle's examples of true actuality. The actuality of the builder is presented here as the exercise of a capacity, but building is a prime example of a movement. Consequently, the exercise of a capacity does not necessarily produce a true actuality, but can also produce a movement. This led Ackrill to conclude that *De Anima* II.5 has little, if anything, in common with *Metaphysics* Θ.6:

> It tells us when to say that a man engaged in an activity is thereby *undergoing* a kinesis, but it does not tell us which of the activities a man may be engaged in are themselves kineseis. If Callias, a skilled man, is building a house, we are not to say that he is being changed (undergoing a kinesis); yet what he is doing, housebuilding, is, of course, a paradigm case of an activity that is a kinesis and not an energeia.[3]

Although Ackrill correctly pointed out that the *kinêsis-energeia* distinction is not at work here, the relation between the two passages can be established, nevertheless. Not every instance of exercising a capacity is an *energeia*, but it is possible to predict just which cases of exercising a capacity will qualify as *energeiai* containing their ends in themselves.

To demonstrate, it is best to begin by seeing how the classification of different types of potentialities cuts across the distinction between motion and actuality. This is achieved by schematizing Aristotle's own examples in 417a30-b28 according to the chart on the following page.

Action	*Agent*	*Patient*	*Nature of action*
housebuilding	builder	building materials	*kinêsis*
thinking	object of thought	thinker	*energeia*
acquiring knowledge	teacher	pupil	*kinêsis*
perception	object of perception	faculty of perception	*energeia*

The capacity to build resides in the agent, and exercising this capacity is a motion; the same applies to teaching. On the other hand, in both cases of *energeia*, the capacity that is exercised is the potentiality of the patient. The question is whether or not the different patterns for activities and motions suggested by the chart hold true in all cases, and in particular, if only those actions are actualities in which the exercised capacity is the potentiality of the patient. That this is true, in general, of motions and actualities, is confirmed by a comparison of *Physics* III.3 and *De Anima* III.2, 425b26-426a26. In *Physics* III.3, Aristotle said that motion always resides in the patient. In *De Anima* III.2, he said the same for sense perception, a prime example of an *energeia*.

In *Physics* III.3, Aristotle said that movement is in the movable, for motion is the actuality of the movable and happens by the agency of what can put it in motion. Nevertheless, motion is the actuality of the agent as well, since the activity of the agent does not differ from that of the patient (202a13-21). Similarly, in *De Anima* III.2, he said that the activity of the perceptible object and perception are one and the same, and that the activity of the perceptible object resides in the patient, just as the activity of what can set something in motion resides in the movable (425b25-426a11). Thus, for both movements and activities, the action resides in the patient. Upon reflection, it is clear that the potentiality of the patient, not the agent, is the decisive factor in determining the nature of the action. The reason for this also accounts for why the action of the agent resides in the patient: because it is the patient that is changed or activated by the action (202a25-31, 426a4-6). Consequently, if exercising the capacity to build can be an *energeia*, then it must be possible to regard the builder as the patient and the exercise of his capacity as caused by an agent.

Ackrill's position would support the claim that exercising the ability to build can be considered an *energeia* when building is regarded as undertaken for its own sake--liberated, as it were, from its involvement with the *telos* of building, and the potential of the buildable to be a house. Hardie argued along these lines when he claimed that the combination of present and perfect tenses does not provide a clear and complete distinction between movements and activities:

> But this criterion will not enable us to separate two kinds of performance since the same performance could be kinetic or nonkinetic according to the way in which it is described.... The criterion would make walking home kinetic but walking, in contradiction of what Aristotle says in the *E. N.*, non-kinetic.[4]

However, taking this position has its liabilities. Hardie's objection does not take account of the fact that Aristotle never considered action verbs, whether they describe movements or activities, apart from some direct object or other complement to the verb. Walking always involves a reference to some distance or destination. This is why we can apply to 'walking' adverbs such as 'quickly' or 'slowly' which imply a relation to distance and time. Activities also imply a direct object, which is closely related to the *telos* of the activity. Penner has shown, with reference to the 'quickly-slowly test, that one may say he 'has enjoyed' a symphony, before the symphony is completed. Still, it is always the *symphony* that one is enjoying at any given point, and even though it may be played quickly or slowly, one does not enjoy it quickly or slowly.[5] If the nature of verbs could be changed simply by considering the action as undertaken for its own sake, there would be no reason to apply criteria concerning the combination of tenses or expressions related to speed.

While Penner recognized the inseparability of verb and complement, he nevertheless suggested that an action, from the point of view of the agent, may be an activity, while from the point of view of the patient, it is a motion:

> Aristotle, I claim, would have accepted that in the case in question there are two actions or action-like entities involved, an actualization (Φ-ing) and a movement (walking) .... In general, if

> there is a natural faculty involved in some movement I carry out, then whether I intend to or not, I am both (a) exercising the faculty and (b) carrying out the associated movement, the latter quickly or slowly, the former not. And one cannot make an informed choice of the one without the other. Thus what we would usually think of as a single action turns out to be two action-like entities.[6]

The question to be considered in the next section is whether there are one or two actions involved in a movement, for Aristotle. After the issues implicated by this question are clear, it will be possible to show how the question is involved in Philoponus' origination of impetus theory and also his theory for the propagation of light.

II

Some of Penner's own remarks militate against the two-entity theory of action he presented in this passage quoted above. If it is true, as Penner said, that the nature of building cannot be changed by considering the action as undertaken for its own sake, the builder can never be a patient who experiences building, the activity, as his actuality. This is because building itself must always resides in the buildable. There must the activity of the builder reside as well. Aristotle's remarks in *De Anima* III.2 support the view that as long as building is movement, exercising the capacity to build must be a movement too. In his remarks about sense perception, Aristotle indicated that although actualization in the case of activity requires an agent, the activity of the agent is itself the realization of the potentiality of the patient:

> For just as both 'acting' and 'being acted upon' reside in the patient, but not in the agent, in this way also the actuality of the perceptible object and that of the faculty of perception reside in the faculty of perception (426a9-11).

Here, Aristotle made clear that, just like a movement, a true *energeia* must reside in the patient. These consideration would imply that Aristotle believed there is only one entity, a motion or a true actuality, involved in any action.

However, to explore the implications of the point Aristotle made here, one must take account of his statements in *Physics* III.3 about the way in which the actuality of agent and patient are the same.

In *Physics* III.3, Aristotle said that the activity of agent and patient are different *tôi eidei* (202b1) and each has a different *ergon* and *telos* (201b24). Nevertheless, one need not say that teaching is learning or that acting is being acted upon in order to say their activity is one (202b10-12). The activity of both is one

> ... just as one in relation to two, and two in relation to one, are the same interval, and the going up and going down; for these are one, although there is not one definition of both (202a18-20).

Aristotle reiterated his point a few lines later. The actualities of agent and patient are one, as the road up and the road down are one, while yet being different, since travelling from Athens to Thebes is not the same as travelling from Thebes to Athens. The analogy suggests that the two aspects of a movement are the same in substratum, since it is the road itself between the two points which remains unchanged, regardless of the direction of travel. But what corresponds to the road in Aristotle's analogy is the motion (202b19-21). Consequently, we learn little more from this analogy than we knew before, namely that Aristotle thinks any motion has two aspects. There is more to learn about the significance of this analogy from 202b5-10, however.

Aristotle, emphasized, in the first place, that the actuality of the agent must reside in the patient. Then, he stated exactly how their activity is one:

> Nor is it paradoxical that the actuality of one is in the other (for teaching is the actuality of the teacher, but in something else, and it is not cut off from that thing but is a this in a this), nor does anything prevent the [actuality] of the two from being one and the same (not by being the same thing, but in the way that what exists potentially is related to its actualizer) (202b5-10).

This passage is a special formulation of a general way Aristotle often used to speak of the similarity or sameness of two things--as potential is related to actual. The principle figures in *De Anima* II.5, where Aristotle spoke of the senses in which *to paschein* is used. In one sense, it means a kind of destruction by a contrary, and in another sense, it is rather a preservation of the potential by what is actual and similar to it as potentiality is related to actuality (417b3-5). The assumption lying behind this and all other uses of the formula is that something existing actually makes the potential *like itself*. All change is due, in one way or another, to this process of actual making potential like itself. In *Physics* III.3, Aristotle cast the formula in terms of agent and patient, *to energoun* and *to dunamei on*, rather than their actualizations, but its import is the same. At the end of *Physics* III.2, Aristotle said that the mover is thought to be a kind of form, which puts another in motion, just like the actual man makes one like himself out of what is only potentially a man. This is only a loose rendering of the principle, and a thoroughgoing application must take into account the goal of motion, since this is the source of a motion's definition. The way in which the motion of agent and patient is the same, in terms of the similarity of potential and actual, can be set out simply in the following way:

patient becoming like agent	=	MOTION	=	agent making patient like itself
patient having become like agent	=	GOAL OF MOTION	=	agent having made patient like itself

The motion which constitutes the activity of agent and patient is the same for both, but what the patient experiences is the inverse of what the agent contributes.[7] Consequently, Aristotle's analogy to the road up and the road down is especially appropriate. As for the issue of definition, it is important to notice the sense in which the goal of motion is identical from either aspect, since the end is achieved when patient is like agent. As the table shows, this same end can be characterized differently. For this reason, the two aspects are different in form and definition. From an objective standpoint, however, there is no difference since the end, for both, is that patient be like agent. This is why Aristotle said that teaching and learning inhere in the same motion (202b21).

It is easy to see how the principle--that actual makes potential like itself--applies to generation and alteration. The way in which the two aspects of motion are one applies especially well to the case of teaching, since the teacher makes the student one who knows a particular subject matter, just as he does himself. But other examples are harder to reconcile to this basic pattern. In housebuilding, for example, the agent does not produce from the building materials another builder like himself. Nevertheless, the principle can apply here as well. Whether the case is teaching or building, a form possessed by the agent brings the potential into full actuality. Aristotle applied the principle to housebuilding in a passage where he considered housebuilding as a case of artificial generation (*Metaphysics* Z.7, 1032b13-14). He said that the craftsman's ability (*hê oikodomikê*) is the form of the house and for this reason, we could say that house is produced from house. The house of the craftsman's knowledge exists without matter, while the house produced on its model has the bricks and other building materials as its matter. It is the "so-called formal nature" which is similar to what exists potentially and which brings it to actuality. The house of the builder's imagination brings into being a house from the building materials. Aristotle identified this model with the craft itself.

If the principle that actual makes potential like itself can be applied with uniformity to all kinds of change, then *Physics* III.3 would appear to provide a sound basis for a one-entity view of motion. The activity of the housebuilder is not separate from the movement of building and not different from it. The key factor in this point of view would be that the activity of the builder must reside in the house being built. The same is true for the activity of any agent that induces motion.

However, the analysis of *Physics* III.3 cannot be decisive for ruling out a two-entity theory, if the proponent of that view can simply claim that *Physics* III.3 applies only to one of the two actions (the motion) that are occurring at the same time. For a one-entity view to prevail, it is necessary to regard any verb as generating only one pattern of relationship among its agent, patient and their respective *telê*. This requirement seems too stringent, since Aristotle's treatment of the agent and patient in different kinds of actions may not reflect an inalterable pattern for each verb, but rather may simply be an artifact of his way of explaining actions. This possibility is brought into relief most clearly by reference to *Metaphysics* Θ.8, one of the few passages where Aristotle compared motion and true actuality without casting each as the actualization of the potentiality of a patient.

He started in this passage with an *energeia* that can mean either the motion or true actuality of *Metaphysics* Θ.6 and proceeded to distinguish the types on two bases: 1) whether or not something is produced in addition to the action (*chrêsis*) (1050a23-29), and 2) where the activity (*energeia*) resides (1050a30-b2). In those actions for which something else comes into existence beyond the action itself, the *energeia* resides in the thing produced, like building in the thing built (a30-34). But in the case of those actions for which there is no other product beside the activity, the *energeia* resides in the thing itself, like seeing in the one who sees (a34-36). In this passage, *energeia* is the activity, whether movement or true actuality, that results from the agent's initiative. According to this formulation, an experienced builder might contemplate the form of a house he expects never even to build. He could undoubtedly conjure up a very complex image, savoring all its fine points of design and detail, in a way no layman could do. In this way, he would be exercising his capacity to build as a true actuality.

The real question, however, is whether or not the builder exercises his capacity as a true actuality at the same time as he is actually building a house. While *Physics* III.3 may not rule out a two-entity view of motion, the range of interpretations under which *Metaphysics* Θ.8 supports it is narrow, because, in that passage, Aristotle treated the activity of the agent as a single action whose type (motion or true actuality) will differ as a function of where the action resides. In the case of vision, the *energeia* is the end, while in the case of housebuilding, the *energeia* is at least more of an end than the potentiality (1050a26-27). This remark is important, because in it, Aristotle identified housebuilding, understood as a motion, with the *energeia* of the agent. Combined with his previous remarks, this means that when the *energeia* of the agent is motion, that *energeia* resides in the patient. The exercise of the builder's craft, therefore, is not an entity separate from the motion resident in the patient.

A two-entity view of action is consistent with *Metaphysics* Θ.8 if the exercise of the builder's craft and building are regarded as two entirely separate and unrelated things. On this interpretation, the builder's craft never issues directly in the movement, building, but functions only as a supervisor of the building activity. While Philoponus held a modified two-entity view of motion, he did not adopt this solution. To examine his interpretation of the way in which the agent's and patient's contributions to a movement are one and his interpretation of the nature of the agent's and patient's respective potentialities, let us begin with his explication of *De Anima* II.5.

III

Philoponus' commentary on *De Anima* II.5 indicates that he was well-aware of the reason Aristotle discussed the different senses of *to paschein* and *alloiôsis*. He stressed the different sorts of potentiality actualized in change, and the way in which the actualization of each involves the destruction of a previous condition. In his commentary on 417b1-5, he stated that the change typified by the exercise of knowledge is not a change as to a certain quality: "For it is not that, when one quality is destroyed, it changes to its contrary, but while the state itself remains, it produces the actuality alone" (*In de An.*, 301.19-21). Acquiring knowledge, on the other hand, is a change in quality, for "what is changing changes from one contrary quality to the other, while the substratum remains the same" (*In de An.*, 301.26-28). He believed that, in 417b12-16, Aristotle had introduced a division within the first type of potentiality. First, Aristotle distinguished between change from one contrary state to another and change from state to activity. Then,

> ... he distinguished two sorts of change in the case of first potentiality--the one, a change into privation, which he called 'being acted upon' and 'being altered' in the proper sense, and the other, a change to the state from privation, which is neither affection nor alteration, or else [those terms are used] homonymously (*In de An.*, 304.34-305.2).

The change to privation involves not only the sorts of change typified by the lapse from knowledge into ignorance, but also change between contraries, because "...there occurs a kind of destruction of that from which the change occurs..." (*In de An.*, 304.23). The change to state from privation, like the acquisition of knowledge, is different, being a change into perfection and "that at which the nature aims" (*In de An.*, 304.25). Thus, Philoponus articulated an understanding of the different senses of *to paschein* and *allōiôsis* not far removed from the modern viewpoint elaborated earlier.

In light of this, it is surprising to discover that he believed that the exercise of a capacity already possessed is always a true actuality of the sort described in *Metaphysics* Θ.6. This view is expressed in his commentary on *De Anima* II.5, 417a14, where Aristotle alluded to the distinction between motion and true actuality without going on to develop it. Philoponus discussed this distinction in relation both to acquiring a capacity and to exercising a capacity already acquired. Motion is the progress from the first potentiality to the state, while activity is the simultaneous actualization of the state:

> Thus, motion is the progress (*proodos*) from the first potentiality[8] to the second of something having to do with substance, when the substance itself is preserved; but actuality is the perfect production (*probolê*) of the state, when the state does not become different (*In de An.*, 296.34-297.2).

This combination lies at the heart of Philoponus' treatment of the scheme of "first and second potentialities" and "first and second actualities." First potentiality is the predisposition (*epitêdeiotês*) of the subject to change in some respect, or to acquire a particular trait. Its actualization is first actuality--the possession of the state (*hexis*). First actuality is also known as second potentiality, because the state itself is only the passive possession of a condition that may be exercised. The exercise of the state is second actuality. The progress from the first potentiality to the second takes time, but the change from second potentiality (or first actuality) to the second actuality takes place in no time whatsoever.

Since Philoponus believed that every exercise of a capacity is a true actuality, how did he deal with the telltale example of the housebuilder? He exercises a capacity when he builds, but the resulting action is a *kinêsis*, not an *energeia*. This example did not assume the same importance for Philoponus that it does for modern commentators. In discussing 417b9, Philoponus said:

> For, as the housebuilder, whenever he is already building, undergoes no alteration as to his craft, similarly neither does the knower, whenever he exercises [his knowledge, undergo an alteration] as to knowledge, nor the subject of perception when he perceives; for it is a change from idleness to activity, not from form to form, which is what alteration effects (*In de An.*, 303.31-35).

The housebuilder can be compared to the thinker because neither changes in form when he exercises his ability. However, the value of the comparison lies in the fact that the builder obviously is not changed when he builds, since he is not the object modified. On the other hand, it is not so obvious that the thinker is not changed, since he is himself the subject in which the action, regardless of its nature, takes place. It would appear that Philoponus understood that the exercise of the builder's capacity results in a motion; this fact is part of the import of the comparison. But this did not prevent his claiming that the builder's exercise of his craft is a true actuality. He used it as one example, along with the geometer's exercise of his knowledge, of the change from state to actuality (*In de An.*, 302.25-28). In this context, it is clear that the builder exercises his craft as a true actuality even while engaged in the process of building (*In de An.*, 301.21-26).

We shall find that Philoponus held a view of motion, according to which an activity that is an *energeia* in the agent may *become* a motion when exercised in a patient outside itself. However, turning to Philoponus' commentary on *Physics* III.3, we find no trace of the sort of two-entity theory of action articulated by Penner and others. In his commentary on *Physics* III.3, 202a25, where Aristotle discussed the *entelecheiai* of the agent and patient involved in a *kinêsis*, Philoponus said that the puzzle Aristotle offered did not concern actualities as wholly different from movements (*In Phys.*, 376.25-27). He allowed only two options: 1) that there are two movements, resident in the mover and moved thing respectively, and 2) that there is only one movement composed of the action of both agent and patient. The latter option is the one he chose. The way in which he limited the options suggests that he would not have accepted a two-entity theory, whereby the same action is a movement from the point of view of the patient and a true actuality from the point of view of the agent. More insight on Philoponus' view of motion and true actuality can be gained by looking at his commentary on *Physics* III.3 in general.

In his introduction to his commentary on *Physics* III.3, Philoponus followed Aristotle closely in listing the objections that might be raised against the idea that there is one actuality for agent and patient. The first objection is that it is impossible for the actuality of one thing to exist in another (*In Phys.*, 372.26). Philoponus established this himself (*In Phys.*, 369.28-29). He then said that it is not impossible but even necessary for things existing in relation that the actuality of one exist in the other (*In Phys.*, 372.26-29). Mover and moved thing are among those things existing in relation, as well as father and son, teacher and pupil, and in general, agent and patient. Another objection is that two things differing in form cannot be the same actuality (*In Phys.*, 373.8-9). Philoponus stated that this too is not impossible for things existing in relation, whenever the one thing exists potentially and the other exists actually, according to this same potentiality (*In Phys.*, 373.10). Then, the actualities of the two are one in substratum, thought not in form. A third problem arises from there being just one actuality for both agent and patient. Teaching would be the same as learning, and the teacher would be the same as the pupil (*In Phys.*, 373.28-30). Although *didaxis* is the same as *mathêsis*, it does not follow that the teacher is also the pupil, since teaching and learning are not the same in every respect but only in substratum. There are two vehicles for the solution to all three problems: the behavior of the category of existence in relation, and the distinction between being the same in substratum and identity in form. As the points made in favor of Aristotle's viewpoint demonstrate, the two types of argument are related.

Philoponus said that the actuality of agent and patient is one in substratum, because of the way potentiality is related to its actuality (*In Phys.*, 373.10-11). He recognized that the substratum in this case is not the moved thing but the motion itself. In his commentary on 202b19, he offered this explanation of how motion is like a substratum in relation to its different aspects:

> Now he says that the process of teaching (*didaxis*) is not entirely the same as the process of learning (*mathêsei*), but that is the same in which teaching and learning inhere, as in a substratum. Just as the ascent and the descent are not the same, but the stairs in which these reside, and so is the motion itself--I mean the walking and putting one foot in front of the other--(for it is properly the same; for it admits of one and the same definition), nevertheless, the ascent and the descent, which inhere in such a motion as in a substratum are not properly the same things; for their definitions are different (*In Phys.*, 383.21-28).

In the introduction to his commentary on *Physics* III.3, Philoponus gave the following examples of the way the actuality of agent and patient are the same in substratum but not in form:

> For the process of whitening (*leukansis*) which the whitening agent effects and the subject of whitening experiences is one, and the subject which the teacher (*ho didaskôn*) teaches and the student learns is one. Also, then, the teaching (*didaxis*) of the subject by which he teaches what (*to didaskon*) the one taught (*to didask omenon*) is taught

> is one. For teaching is nothing else but learning from a teacher (*to ek tou didaskalou manthanein*). And similarly, also, for learning (*mathêsis*) (*In Phys.*, 373.12-16).

The significance of these examples lies in their relation to Philoponus' distinction between two forms of the verbal noun: *leukansis* and *to leukanein*, *didaxis* and *to didaskein*, *mathêsis* and *to manthanein*. While *didaxis* and *mathêsis* may be the same, *to didaskein* and *to manthanein* are not. He explained in more than one passage why he distinguished between the two descriptions of the same process:

> Whenever I say process of whitening (*leukansin*) or process of teaching (*didaxin*), I mean the actuality itself, but whenever [I say] whitening (*leukanein*) or teaching (*didaskein*), [I mean] the being moved and actualized according to actuality or movement (*In Phys.*, 371.31-372.1).

> For the process of teaching (*didaxis*) manifests motion itself and actuality in itself, but teaching (*to didaskein*) [manifests] the relation of the actuality of the teacher to the pupil (*In Phys.*, 380.11-13).

When he referred to movement and actuality in these passages, Philoponus was not distinguishing between movement and true actuality but was emphasizing the nature of motion as a kind of actuality. In this neutral meaning, *energeia* describes the deed which is the substratum for the different forms of the same motion. *Didaxis* is the activity itself, while *to didaskein* is the activity of the two things which stand in relation to one another, the teacher and his student. The reason for distinguishing these forms of the verbal noun so carefully is Philoponus' belief that *didaxis* and *mathêsis* are the same process described under different relations (*In Phys.*, 381.18). Emphasizing the sense in which *energeia* may apply to movement creates a neutral sense of the term that applies to both process verbs at once. The belief that teaching and learning are the same is integrally related to Philoponus two-entity view of motion. This is clear from Philoponus' consideration of the exercise of the teacher's capacity even when the student does not learn.

In commenting on Aristotle's statement that the *energeia* of the teacher is not separated from the pupil (202b7), Philoponus said one must not regard the actuality of the teacher as simply the production of thoughts. For even when he is not teaching but is alone, he manipulates thoughts. He continued:

> As even if, when the student is present, the teacher speaks without affecting the student, it is not said that he teaches, in this way, clearly, the process of teaching (*didaxis*) is not simply the articulation (*prosbolê*) of thoughts but the inducement (*peithô*) generated in the student by the words of the teacher which is called the process of learning (*mathêsis*) by another relation (*In Phys.*, 381.15-18).

The relation between the knower's thinking his thoughts and teaching is treated again in a passage where Philoponus presented the theory of imparted power he used to explain causation in the physical world. In this passage, the relation of the thinker and the student is explicitly tied to the unity of the motion in which agent and patient are involved:

> The knower produces thoughts, even when there is no one to learn, but whenever this production, happening in a student, acts (*drasêi*) in the student, this active power (*drastêrion tautên dunamin*) by which the student is affected, this I say is one and the same ... (*In Phys.*, 385.4-8).

This passage differs slightly from Philoponus' earlier formulation. Philoponus here said that the same capacity which would be a true actuality when exercised in isolation becomes a movement when expended in a patient other than the subject possessing the capacity. The exercise of knowledge becomes a motion, when the knower directs the exercise of his knowledge toward making another a possessor of knowledge like himself. Knowing becomes teaching.

Philoponus affirmed Aristotle's statement that the *energeia* of the teacher is not separated (*apotetmêtai*) from the pupil, adding that the *energeia* comes from the teacher but is exercised in the pupil. Furthermore

> it is necessary to conceive the actuality of the teacher not simply as the articulation of thoughts (for even when not teaching but being by himself, he thinks (*procheirizetai*) thoughts), but as the motion generated in the soul of the student by thoughts (*In Phys.*, 381.11-15).

Philoponus' view of motion articulated here has much in common with Aristotle's approach to the *kinêsis-energeia* distinction in *Metaphysics* Θ.8. Philoponus emphasized the location of the agent's activity in the patient. Furthermore, he pointed out that when the activity is exercised in the thing itself, it is a true actuality, while its location in the patient indicates the action is a motion. Philoponus went beyond Aristotle, however, in saying that the same thing which is an *energeia* in the agent becomes a motion when expended in another. This entity can be identified whether its possessor is exercising his knowledge or teaching. This is one of the new dimensions of the *energeia* concept introduced by Philoponus, and it represents his solution to the problem of the unity of the agent's and patient's contributions to the same movement.

Philoponus presented this example of the thinker becoming teacher to bolster his assertion that something is transmitted from mover to moved thing in locomotion. Philoponus presented this idea of imparted power to explain how the motion of agent and patient is one, even when the mover moves another because it is in motion itself. He said that all physical things are moved in this way.[9] Concerning the imparted power, he said:

> The motive power (*kinêtikê dunamis*) endowed by the mover in the moved thing--I mean this force (*ischus*) and the actuality according to which also it is moved, the potentially movable being perfected by such an actuality--[this motive power], being one and the same, has its principle of procession (*proodou*) from the mover, but its end and accomplishment and its 'remaining'(*monên*), as it were, in the moved thing. For, being generated in the potential, it is not itself expended, but, as it were, 'remaining' (*menousa*), it comes to perfection, and this perfection is the advancement (*proagôgê*) of potentiality, and this is motion (*In Phys.*, 384.33-385.4).

The *drastêrios dunamis*, which is teaching when actualized in the student, is comparable to the *kinêtikê dunamis*, which is the same for mover and moved thing. The motive power imparted to the patient is similar to the actuality which becomes a motion in the case of teaching.

It is important to notice that the *drastêrios dunamis* is not by itself the *energeia*, which is knowing in the agent or teaching when expended in the patient. The *drastêrios dunamis* is the contribution of the agent which requires the potentially movable in the patient to become motion. *Dunamis*, used in this way, is not a potentiality in the classical Aristotelian sense, because this *dunamis* actualizes a potentiality in the patient. Rather, *dunamis*, used of the agent's contribution, is closely related to *energeia*, the difference being that the *dunamis* must actualize a potentiality in another before it is effective. The *kinêtikê dunamis* completes the potentially movable in the subject.

The concept of *energeia* has also subtly changed in Philoponus' account of motion. Let us reconsider the builder, exercising his capacity as a true actuality, while yet engaged in a movement. Philoponus said that an activity which is a true actuality becomes a movement when exercised in another as patient. However, according to Philoponus' treatment of the builder in *De Anima* II.5, the agent exercises his craft as a true actuality, even if the agent produces motion in another. The conclusion must be that the actuality of the agent which is contributed to a motion like housebuilding is not the whole of the agent's actuality. This conclusion is consistent with the passage where Philoponus presented his theory of imparted power. The actuality contributed to the patient is one with the potentially movable of the patient. Knowing becomes teaching, but there is no indication that the knower's exercise of his knowledge consists completely in this contribution. Indeed, considering the nature of true actuality, it would appear that during the time of the exercise of second actuality, there is an inexhaustible supply that may be converted to motion when exercised in an appropriate patient. This is because true actuality is indivisible in form and requires nothing to complete it. If a part can be contributed at all to another, then what is left must be just as complete as it was before that part was removed.

Before returning to the propagation of light we must see how Philoponus' theory of imparted power is related to the definition of motion. Philoponus considered Aristotle's definition of motion first in its form "motion is the actuality (*entelecheia*) of the potential as such." He explained the last phrase of the definition in the following way:

> He added the "is as such"--that it, "is this particular thing *qua* potentially"--in order that the actuality might come to be, while the potentiality of which it is the actuality, still remains and is preserved (*In Phys.*, 342.15-17).

Entelecheia, for every thing, has two meanings. There is the *entelecheia* of the perfected thing when it is no longer potential, and the *entelecheia* of the thing when it is being moved and led to its form (*In Phys.*, 342.17-20). Philoponus did not maintain that there are two kinds of potentiality involved in the same motion, one for motion and one for the perfected state. There is only the one potentiality for becoming this or that particular thing. There are, however, two forms of actuality, one which preserves the potentiality and the other which eliminates it by bringing into actuality its corresponding form. Nevertheless, Philoponus distinguished between *to kinêton* and *to dunamei* of the same thing. He explained this in his commentary on what he considered to be Aristotle's more precise definition of motion. He believed that defining motion as the actuality of the movable *qua* movable is better than defining it as the actuality of the potential as such. The former definition is superior because it avoids any possible confusion between motion and true actuality:

> ... it is reasonable that Aristotle adopts this definition, also because, as I already said, being led from potentiality to actuality is said not only of imperfect actuality--I mean [the change] from the first potentiality to first actuality--but also of [the change] from the second potentiality to the second actuality (*In Phys.*, 368.30-369.5).

There are two different kinds of potentiality involved here, but only one of them has to do with motion. Philoponus related the definition of motion to the scheme of first and second potentialities and actualities. When Philoponus commended the second definition of motion as more exact, he meant simply that it distinguishes between the first and second potentialities of his scheme. He considered 'the movable' as synonymous with 'first potentiality.' He agreed with Aristotle that the movable is not a special kind of potentiality whose actualization is motion. However, he thought that motion, and not first actuality, preserves and perfects first potentiality, because potentiality and the form succeeding it cannot coexist. This led him to maintain that as long as the potentiality persists, motion continues. The potentiality is gradually depleted as progress is made toward the final goal (*In Phys.*, 366.25-26). In his own theory of projectile motion, Philoponus presented this same idea in terms of the dissipation of the power imparted to the projectile (*In Phys.*, 644.17-19). He argued that the projectile would lose its impetus, whether it travelled in a plenum or a void. The loss of the imparted power reflects the fact that motion actualizes potentiality, and that when the end is achieved, the potentiality has been completely eliminated.

IV

Philoponus' doctrine of imparted power is the vehicle for a neutral sense of *energeia*, which serves his explanation of both motion and true actuality. In its neutral sense, *energeia* is equivalent to the effective power transmitted from agent to patient whenever something is actualized. As such, it also coincides with the *phôtistikê dunamis* of Philoponus' propagation of light theory. The potentially transparent is an example of a capacity possessed by the patient which is exercised due to the influence of an agent, the luminous object. Light is an *energeia*, because it is the activation of a quality already possessed by the medium. Once the part of the medium nearest the luminous body has received the *phôtistikê dunamis*, that part of the medium is capable of transferring the same power to its neighbor. This transmission of the power and the production of the effect, namely light, occur all at once and without the passage of time, because each bit of the medium possesses the trait of transparency as a state whose actualization is a change from first actuality (or second potentiality) to second actuality. For this reason, the actualization of light produces a true actuality, not a movement, and what is described as a succession of effects in fact takes place simultaneously.

This much, then, of Philoponus' theory for the propagation of light can be explained in terms of Philoponus' response to a set of problems in Aristotle's theory of motion. However, the luminous power possesses a tendency to replicate itself indefinitely. As we shall see in the following chapter, this aspect of the theory of light is due to certain modifications of the *energeia* concept of *Metaphysics* Θ.6 that result from its application in Neoplatonism.

Notes

1. See Michael Wolff, *Fallgesetz und Massebegriff*, 88-99, and Nelly Tsouyopoulos, "Die Entstehung Physikalischer Terminologie aus der Neoplatonischen Metaphysik," *Archiv für Begriffsgeschichte*, 13, No.1 (1969): 26-28.

2. For insight on this point, I am indebted to an unpublished paper by John Cooper, "Aristotle on the Ontology of the Senses," pp. 17-18 and n.21.

3. Ackrill, "*Energeia* and *Kinêsis*," 141.

4. W.F.R. Hardie, *Aristotle's Ethical Theory* (Oxford: Clarendon Press, 1968), 308.

5. Penner, "Identity of Actions," 414-420.

6. Penner, "Identity of Actions," 439-440.

7. Mary Louise Gill treated this point in more detail in "Aristotle's Theory of Causal Action in *Physics* III.3," *Phronesis* 25 (1980): 136.

8. I use the terms 'potentiality' and 'actuality' for *to dunamei* and *to energei* respectively, to make for easier reading. Philoponus used both the terms *dunamis* and *to dunamei*, as well as *energeia* and *to energei* in his discussion of first and second potentiality and actuality.

9. Philoponus thought it was clear that, for alteration or growth, the movement involving both agent and patient resides in the patient alone, since the agent itself does not change. However, he saw a problem when the doctrine is considered in terms of locomotion, for the mover may move another because it is itself in motion (*In Phys.*, 384.13-21). In this case, there are two separate motions for mover and moved thing. For example, the sphere of fire around the earth is moved by the heavens, but the movement of the heavens is separate from that of the sphere of fire, both in definition and in substratum. This may seem a strange interpretation of Aristotle's doctrine in *Physics* III.3, because Aristotle was concerned only with the unity of the actualizations of agent and patient for a single motion resident in the patient. Philoponus considered the problem of the unity of two

separate motions, where the patient in one movement is the agent of movement in the other. However, Philoponus regarded this neither as a separate problem or a special case of the larger doctrine, for he maintained that all things in the physical world involved in locomotion, are moved in this way (*In Phys.*, 384.31-33). Philoponus solved the problem of the unity of the two movements by maintaining that a motive power is endowed in the moved thing by the mover. It is this motive power which is one and the same in mover and moved thing, and the accomplishment or completion of this power is motion itself.

Chapter Five

Reexamining the Case for Neoplatonic Influence on Philoponus' Causal Theories

I

Many of Philoponus' innovations in the theory of physical causation can be explained in terms of his response to problems in Aristotle, without resort to Procline influence. There are Neoplatonic elements in Philoponus' natural philosophy, but determining the nature of the Neoplatonic influence can be complicated because Proclus used so many Aristotelian concepts in his own system. In his commentaries, Philoponus used many Procline terms for concepts that originated in Aristotle. This indicates how thoroughly Neoplatonism permeated philosophical thinking in the sixth century. However, the use of terminology alone does not inform us that a Neoplatonic solution has been enforced to any given problem. For that, it is necessary to determine what elements in Philoponus' analysis could not have originated in Aristotelian concepts and problems. Bearing this in mind, we should examine the use Proclus made of the *kinêsis-energeia* distinction. Many of the similarities between the causal theories of Proclus and Philoponus--including the doctrine of imparted power--are due to the implications of this Aristotelian concept, used by both philosophers. Nevertheless, certain elements in Philoponus' approach to natural causation are specifically Neoplatonic in character. Upon completion of this examination, the exact nature of the Neoplatonic contribution should be clear.

In the participation theory of Procline Neoplatonism, a cause produces an effect, simply by virtue of its own being (*autôi tôi einai*). It possesses a superfluity of power which results in the spontaneous production of a host of effects. Scholars have accounted for this active power attributed to the cause, in a variety of ways. Dodds traced its history back through Plotinus to the Stoics.[1] Trouillard explained it in terms of the necessities of Proclus' system: for the creator, the production of the universe can involve no act of will or diminution of his own being, either of which involves a change in the substance of the creator. This is impossible, since the creator is already complete in every way. Consequently, he creates simply by virtue of his existence. What applies to the creator in relation to his creation applies as well to any cause in relation to its inferiors.[2] More recently, Gersch understood the superfluity of power in terms of its infinity or inexhaustibility, which is a characteristic of the spiritual world.[3] Each of these accounts has something to commend it, but none takes account of the strong Aristotelian contribution to the relation between cause and effect. The idea that effects are logically necessary products of certain existents derives from the identification of members of the causal chain in participation theory as cases of Aristotle's true *energeia*. The importance of the *kinêsis-energeia* distinction to Proclus' participation theory is most clearly seen in certain passages from his commentaries on the *Parmenides* and the *Timaeus*. However, once its role has been pointed out, it is easy to see it at work in the *Elements of Theology* as well.

In the case of motion, it takes time for the end to be achieved. For true actuality, the end is present from the very beginning of the action. Turning to Proclus, we can see he made use of this contrast. In explaining how the universe is generable while yet eternal, he invoked the combination of present and perfect tenses which is such an important part of Aristotle's original characterization of true *energeia*. In his commentary on *Timaeus* 28b-c, Proclus established that the universe is not generable, in the sense of having a beginning in time, nor simply because it derives from a cause.[4] The universe is generable in that it is always at the same time coming to be and has come to be (*ôs aei gignomenon hama kai gegenêmenon*) (*In Tim.*, I.290.24). The universe is still coming to be when already in existence, because it is continually reliant on its cause. Proclus expressed this by comparison to the reliance of light on a continuous light source:

> For as sunlight always proceeds from its own source, similarly also the universe always comes to be and always is being created

> (*paragetai*), and is always at the same time coming to be and has come to be (*In Tim.*, I.290.29-291.1).

Light proceeding from the sun would disappear without the continual presence of the sun. In the same way, the universe depends on its cause for existence. However, in the case of the universe, the combination of present and perfect tenses indicates a permanent and unending relation between cause and effect. In Chapter Three, it was pointed out that what is remarkable about true actuality is that the present tense can still be used when the perfect applies, i.e., the action continues while the end is present. Proclus' use of the present and perfect tenses capitalizes on this facet of *energeia*. A composite thing, even when it has been combined, is still being held together (*In Tim.*, I.291.3-4). In a similar way, the universe is coming to be and has come to be.

While we can make this correlation between Aristotle's and Proclus' uses of the combination of tenses, this is not the only indication that Proclus used Aristotle's concept. Proclus came much closer to both sides of Aristotle's contrast between motion and true actuality in his distinction between *energeia atelês* and *teleia kinêsis*. In his commentary on Plato's *Parmenides*, Proclus discussed the way in which motion resides in *Nous*. There can be nothing in the effect which does not exist in a more perfect form in the cause. Since there is motion in the universe, so also is there motion in its cause, *Nous*. The motion in *Nous* is of a kind different from the motion in the universe, because whatever is changeable has parts, and *Nous* is simple and ungenerable. This more perfect motion residing in the cause of the universe consists in *Nous* thinking itself:

> If the Demiurge exists, and if it thinks itself, it is evident that it is being actualized. Indeed, actuality is called perfect motion, and of such a kind is the creative agency (*dêmiourgikê poiêsis*) and thought.[5]

This perfect motion is the production of the cosmos, insofar as it is immaterial, i.e., the production of the form of the universe and the tremendous variety of forms contained in it. Later, we shall explore the idea of *Nous* thinking itself in relation to the creation of the universe. For the moment, however, let us investigate the difference between the two types of motion described by Proclus.

Proclus contrasted the production of the universe of forms to the debased form of motion present in the material world:

> And [Nous] possesses the motion of the mover, and body the [motion] of the movable; and the motion in the latter is the image (*eidôlon*) of the motion pre-existing in the former. For the motion of the former is perfect motion, for it is actuality. But the motion in the latter is imperfect actuality; and the imperfect has its existence and completion from (*ek*) the perfect (*In Parm.*, 797.32-38).

The reference to *energeia* which is *atelês* recalls one of Aristotle's descriptions of motion--as a kind of actuality, but incomplete (*Physics* III.2, 201b31). Its companion in the Procline passage, *teleia kinêsis*, is not a term of Aristotle. However, this phrase represents the actuality which is Aristotle's alternative to motion. This is not immediately apparent, until we notice the time reference Proclus used to describe the relation between cause and effect. He said that the Demiurge is cause *pantôn athroôs* (*In Parm.*, 771.4). Although the material universe is immersed in time, there are causes in the material world which act in a way similar to the Demiurge. These causes require no time to produce their effects. Fire conveys heat to others, the sun produces light, and the soul life, in the same way that the Demiurge produces its effects (*In Parm.*, 772.1-8). In the commentary on the *Timaeus*, Proclus described the procession of light from the sun in a similar way, as an *achronos hupostasis*. He also described the generation of the universe as an *apogenêsis* which is *achronos* (*In Tim.*, I.84.2, III.32.2). As we have seen, the Aristotelian commentators used this same time reference, *athroon* or *achronôs*, to describe the transition from potentiality to actuality, when they distinguished true actuality from movement. This time reference acquired an additional meaning for Proclus that it did not have for the Aristotelian commentators. For them, *achronôs* implied that though the action had a beginning, it need not cease. For Proclus, the time reference implied not only that the action does not cease, but also that it has no beginning. The most important aspect of the *energeia* concept remained unchanged, however. This is the presence, at every stage of the action, of the end for which the action was undertaken.

Citing Porphyry, Proclus explained the production of the universe by the Demiurge by contrasting it to the work of a craftsman who needs tools to implant the form in his material. By planing and chiseling, he removes everything from the matter which will not receive the form. However, if there were no impediment in the matter, the form would be produced in the substratum in no time whatsoever (*In Tim.*, I.395.10-22). This is precisely the situation in the case of *Nous* thinking itself. In contrast to the craftsman who must work with a recalcitrant matter, there is no impediment to the creative activity of the Demiurge. For Aristotle, the activity of mind was a paradigm case of true actuality. He regarded thinking as a verb which describes an action containing its end in itself. In the activity of thinking, the objects of thought are present in the imagination, as soon as the mind directs itself to them. Proclus' description of *Nous* is based on this model, but the fact that the activity of thought is reflexive has special results. It means that *Nous* creates the universe of forms. The inevitable production of other existing things is not ordinarily the result of the exercise of true actuality. It happens in this case because *Nous* exists and its essence resides in thinking itself. Consequently, the thought of *Nous* must exist, as surely as it exists. We are left with the fact that the products of *Nous* are a necessary part of its actuality. Thus, creation is as much the essence of *Nous* as thinking.[6]

Because a cause entails its effects, there is a sense in which cause and effect constitute one entity. Proclus expressed this unity with his term *kuklikê energeia*. He used this term to characterize the whole cycle of movement of an effect from its cause and back to its cause. The triad, *monê-proodos-epistrophê*, describes the structure of this cyclic movement. The first stage, *monê*, expresses the fact that every cause produces its effects, while itself remaining steadfast (*El. Th.*, prop. 26). The cause remains unchanged and undiminished by its activity of producing others. At the second stage of the cycle, *proodos*, the cause proceeds into generation of what it is capable of producing (*El. Th.*, prop. 25). This stage reflects the necessity that some part of the cause become separate from it in order to produce another. From being separate, it becomes different by actualizing a potentiality inferior to that of the cause. When this inferior potentiality is actualized, it turns back to its cause. In this third stage, *epistrophê*, the effect and its cause become one in some way. It might be more exact to say the effect imitates the cause as closely as possible, given its inferior character. This triadic structure, which constitutes the *kuklikê energeia*, represents an 'exploded view' of the logical implications of the *energeia* concept, as Proclus used it. The procession stage of the cycle represents the bond of entailment connecting cause and effect. It resembles very closely the *drastêrios dunamis* of Philoponus' commentary on *Physics* III.3.

Proclus said that in every relation between cause and effect, it is necessary that something pass over (*pareinai*) from cause to effect (*El. Th.*, prop. 12). This element Proclus called *dunamis*, *dosis*, or in language identical to that of Philoponus, *drastêrios dunamis* (*In Parm.*, 772.12). The potentiality (*dunamis*) which remains in the cause and the one which returns to the cause each has a substratum. They remain separate, in spite of the 'effort' of the effect to imitate the cause as closely as possible. The power (*dunamis*) which proceeds from the cause makes the *ateleia dunamis* of the effect a *teleia dunamis* with respect to what succeeds it. This power is not a combination of matter and form, like the *teleia dunamis*. It is rather the actualizer of the pure potentiality of the patient. In this sense, it is an active power. Once combined with the potentiality of the patient, it becomes a *teleia dunamis*. The active power proceeding from the perfect potentiality does not change into incomplete potentiality, but makes this incomplete potentiality, which is already resident in the patient, as perfect as possible. Thus, the *dunamis* of procession is different from the *dunamis* of the cause or the effect. The *ateleia dunamis* of the effect is Aristotle's potentiality. The *teleia dunamis* is Aristotle's true *energeia*, as an actualized potentiality. The *dunamis* of procession may be understood in terms of the Aristotelian framework for change, which is common to motion and true actuality. An agent acts on a patient which is potentially like the agent. The *dosis* is the agent's contribution to the actualization of the patient.

II

Proclus used the concept of true actuality in an Aristotelian framework for change. Bearing this in mind, let us return to Philoponus' commentary on *Physics* III.3 and his concept of *kinêtikê dunamis*, and inquire to what extent his analysis of change relies on Proclus. He and Proclus used similar terms to describe the force imparted to the patient. In explaining how the actuality of agent and patient are one in movement, Philoponus used the Neoplatonic language of procession and subsistence: "... so that it is not impossible that there is one actuality of both, proceeding (*proiousan*) from the agent but subsisting (*huphistamenên*) in the patient" (*In Phys.*, 374.21-22). In his commentary on 202b21, a passage examined in Chapter Four,

Philoponus said that the imparted power

> ... has its principle of procession from the mover, but its end and accomplishment and its 'remaining,' as it were, in the moved thing. For, being produced in the potential, it is not itself expended, but as if remaining, it perfects, and this perfection is the advancement of potentiality, and this is motion (*In Phys.*, 385.1-4).

The juxtaposition of 'procession' and 'remaining' suggests that Philoponus meant to invoke the causal structure familiar to Neoplatonism, wherein the effect, upon returning, is said to remain in the cause. Michael Wolff interpreted the passage this way, asserting that Philoponus used the Neoplatonic meaning of *monê*, the only difference being that while Proclus made the effect remain in the cause, Philoponus had the cause, in the form of the imparted power, remain in the effect.[7] However, this would constitute a rather important change in the meaning of *monê*. The 'remaining' of the effect in the cause, for Proclus, indicates the continual dependence of the effect on its cause. In the same way, the 'remaining' of the cause in itself refers to the unchanged state of the cause while being the continual source of existence for its products. These two sorts of 'remaining' thus share a common meaning and reinforce one another. Philoponus' use of 'remaining' refers only to the way a cause can continue to act in an effect, by being resident in it. The dimension of continual dependence has been lost. In fact, the imparted power ceases to reside in the effect as soon as the potentiality, whose actualization results in motion, has been exhausted. Consequently, if this is the Neoplatonic concept of 'remaining,' one of the most important connotations of that terms has been lost in Philoponus' application. And one important aspect of the meaning of Philoponus' use of the term has not been associated at all with Proclus' meaning of the term. This aspect involves the relevance of the term *monê* to Philoponus' interpretation of natural motion and natural place. The use of 'remaining' in this context appears in Philoponus' commentary on *Physics* III.5.

In *Physics* III.5, 205b1-10, Aristotle criticized Anaxagoras for his explanation of why the infinite remains stationary. Anaxagoras said that the infinite maintains itself fixed. It is in itself and nothing else contains it. Where any thing is, there it is its nature to be. But Aristotle argued that something might be held in its place by force. Something might be where it is because there is no other place to which to move. In that case, its immobility is not due to a natural disinclination to move, but simply to its being prevented from moving. In his commentary on this passage (*In Phys.*, 448.13-450.27). Philoponus affirmed this criticism. He said that the real reason for something remaining stationary must explain "why, when there is place in every direction, nevertheless it does not move at all" (*In Phys.*, 449.16-17). In reference to the earth, he said that since it has some place to which to move but does not, the cause of its remaining is a natural inclination to be in the middle (17-19). To say that something keeps itself fixed is not to give a natural reason for motionlessness, because it involves the notion of force. But remaining is the opposite of the forced (*biaios*) (*In Phys.*, 449.25-450.1). This sort of remaining is the function of any body in its natural place. As we shall see, it can also be a function of an imparted power behaving in accordance with nature.

Philoponus spoke of earth's inclination to remain in the middle in relation to the formation of natural places (*In Phys.*, 633.19-30). He said the differences between up and down is not due to differences within space itself, but rather to the fact that the bodies of the cosmos assume a particular relation to one another. The part which occupies the middle region we call down, and what occupies the outer extreme up. These regions are the homes of the elementary bodies of earth and fire, respectively.

The denial of natural place as a function of space implies that a clod, when it falls, does not seek a downward place, but rather strives to regain its whole, the body of earth located in the center (*In Phys.*, 439.13-15). This striving reflects its desire (*ephesis*) to maintain the order (*taxis*) established by the Demiurge (*In Phys.*, 581.20). This order is the relation of all the elementary bodies to one another in their formation of the cosmos (*In Phys.*, 633.5-13). The movement of an element to regain its whole constitutes natural motion in the proper sense: the movement of an elementary body when not compelled in some direction contrary to its nature (*In Phys.*, 797.21-22). From this point of view, natural motion and the maintenance of the elementary masses in a particular region are aspects of a single tendency on the part of the elementary bodies.

Strictly speaking, no movement is natural, because nature is the principle of motion and rest, and movements are not causes of motion (*In Phys.*, 200.4-5). But insofar as movements, which Philoponus called *teleiôtikai energeiai*, can be natural, those which proceed from an internal principle merit the designation *to phusei* more than those induced by an outside cause. These latter are only in accordance with nature (*kata phusin*) (*In Phys.*, 200.12-16). Philoponus believed projectile motion is in accordance with nature. In this, he differed from Aristotle, who regarded projectile motion as contrary to nature and due to an outside cause. Philoponus believed projectile motion is due to an imparted power. To this extent, it is no longer due to the agency of another (*In Phys.*, 875.25-27). The projectile has acquired, albeit artificially, an internal principle of motion. For this reason, it resembles *to phusei.*

These remarks of Philoponus suggest that the theory of imparted power takes natural place and motion toward natural place as its model. Philoponus said the imparted power has its completion and "something like remaining" in the moved thing. The significance of its remaining in the moved thing is that the power continues to produce motion as long as some part of the object's potentiality remains unactualized. Once something exists in actuality, its potentiality has been destroyed. But while that potentiality persists, motion is its actualization and preservation. The identification of 'perfecting' with 'preserving' in Philoponus' analysis of motion makes clear the similarity between the remaining of earth at the center of the cosmos and the residence of the imparted power in the moved thing. The imparted power preserves the potentiality of the moved thing, just as an elementary mass maintains its place in relation to the other masses.

If the *monê* used in connection with the continuing presence of the imparted power in the moved thing does not carry a Neoplatonic meaning, what is the Neoplatonic contribution to Philoponus' theory of motion? It is the notion of superfluity of power, which became part of the Aristotelian concept of *energeia.* Because of this superfluity of power, a succession of effects can be sustained. One thing, once it is actualized, is also capable of actualizing another in a similar way. In Philoponus' explanation of the propagation of light, each element in the succession of effects possesses a superfluity of power. The sun, for example, acts in the air immediately touching it and makes it actually transparent. This part of the air, having become transparent, "possesses the power so that it also acts on the things nearby and makes them of this same sort, and in this way, the illuminating power proceeds out to the limits..." (*In de An.*, 330.8-11). For each part of the medium that receives light, actualization brings with it the power to act on others. The idea that an effect not only can but must act as a cause arises as a result of Proclus' use of the concept of true actuality in relation to the activity of *Nous. Nous* thinking itself entails the existence of a whole range of effects. This entailment is at the root of the superfluity of power possessed by a cause, because each effect imitates *Nous* by producing others in a similar way. Proclus explained the relation between each cause and effect in terms of an element passed over from one to the other. This element is the same one to emerge in the commentary on *Physics* III.3 as the actuality that is one for both agent and patient. This is *energeia* in the neutral sense, i.e., neither movement nor true actuality. It corresponds to the *phôtistikê dunamis* in the theory of light, and the *kinêtikê dunamis* in the explanation of projectile motion. However, the superfluity of power possessed by a cause is the result of applying the *energeia* concept to the special case of *Nous* thinking itself.

There are no doctrines or concepts of Aristotle that could have directly given rise to this notion of the superfluity of power. There is one Aristotelian passage, however, which should not be overlooked in discussing the origin of the concept of imparted power. This passage contains Aristotle's tentative suggestion for explaining how a projectile remains in motion after receiving its initial impetus (*Physics* VIII.10, 266b28-267a13). When the agent originally propels an object, he also sets in motion the air around it, and this air then serves as a mover. But there is no more reason for the air to keep moving after the initial impact than for the projectile itself. If the air is to serve as the agent after the projectile has left the hand of the thrower, then we must assume that the air does not cease acting as a mover at the same time it ceases moving (267a5). It stops moving, once its mover has ceased moving it. But it is still a mover and so it moves something next to it (a6-7). Contiguous or successive parts of the air comprise a succession of movers which keep the projectile in motion. The succession comes to an end because the power (*dunamis*) of the mover transmitted in each successive part is continually diminished. The motion ceases when the power transmitted makes something move, without making it a mover as well (267a8-10). Aristotle here recognized the power to cause movement as something which is transmitted from one thing to another, along with movement. He even used *dunamis*, in its active sense, to describe this power. In addition, it is clear that this power is separable from movement, because it can still serve to move another after the thing in which it resides has come to rest. The ability of a moving thing to cause movement in another, as Aristotle characterized it in this passage, harmonizes well with Philoponus' concept of the *kinêtikê dunamis*, which is transmitted from one thing to another. However, Aristotle applied this disjunction of moving and being a mover only to projectile motion. His explanation is tentative and bears little relation to his larger doctrine of motion. Furthermore, he used this *ad hoc* explanation to justify how *air* moves the projectile. But Philoponus rejected the idea that air contributes to projectile motion (*In Phys.*, 642.3-6). Finally, there is no indication that

Philoponus believed a moving thing retains its power to move another, after its own motion has ceased. It does not appear that Philoponus took sufficient note of the finer points of Aristotle's remarks on projectile motion in this passage to justify ascribing to them a large role in the formation of his own ideas on imparted power. We must look beyond Aristotle's explanation of a succession of movers for the source of Philoponus' active power.

There is a strong affinity between Philoponus' uses of the term *energeia* and Proclus' uses of the same term in his participation theory. This is not surprising, once we recognize that Proclus also made use of Aristotle's distinction between motion and true actuality. We have seen how Proclus likened the activity of *Nous* to certain cases of natural causation: the production of warmth in the air by a heating source, and the production of light in the air by the sun or another luminous body. These are precisely the cases to which Philoponus assigned his own succession of effects. In his analysis of these cases, Philoponus relied on a concept of true actuality that he shared with Proclus as a legacy from Aristotle. However, his use of *energeia* is particularly Neoplatonic in character, to the extent that actuality tends to replicate itself in an immediately adjacent substratum which is capable of receiving it.

Notes

1. Proclus, *Elements of Theology*, 2nd ed., ed. with trans. and commentary by E.R. Dodds (Oxford: Clarendon Press, 1963), 212-14.

2. Jean Trouillard, "«Agir par son être même»: La causalité selon Proclus," *Revue des Sciences Religieuses* 32 (1958): 350-51; see also Dodds, ed., *Elements of Theology*, 204-5.

3. Stephen Gersch, *Kinêsis Akinêton: A Study of Spritual Motion in the Philosophy of Proclus* (Leiden: E.J. Brill, 1973), 29.

4. Proclus, *In Platonis Timaeum Commentaria*, ed. E. Diehl (Leipzig: B.G. Teubner, 1903-6), I: 290.17-23.

5. Proclus, *In Platonis Parmenidem*, in *Opera Inedita*, compiled and emended by V. Cousin (Paris: A. Durand, 1864), 771.26-30.

6. Proclus did not state that the effects are of themselves the *telos* of cause. This would imply that they are required by the One for its own completion. However, he did state, in his discussion of the participated intelligences in the *Elements of Theology*, that what acts without motion always has its existence in production (Dodds, ed., prop. 174). This is as close as he came to maintaining that the production of effects is part of the *teleiôtês* of any *teleia dunamis*.

7. Wolff, *Fallgesetz und Massebegriff*, 92.

Chapter Six

The Physical Basis for Mathematical Demonstration in Optics

I

When Philoponus presented his reconciliation of the mathematical and philosophical aspects of optics, he said that it makes no difference, in terms of geometrical optics, whether rays proceed from the eye to the object or *vice versa*. What view of the nature of optics and its relation to physics lies behind this statement? A full answer to this question will involve Philoponus' treatment of optics as a subalternate science, in his commentary on the *Posterior Analytics*. This part of the answer is the subject of the next chapter. One part of the answer, however, must address the physical basis for mathematical demonstration in optics. If the visual rays of geometrical optics have no counterpart in an extromission theory of vision, what are we to think of the directionless mathematical lines Philoponus believed are compatible with either an intromission or extromission theory of vision? Are we to suppose anything in nature corresponds to them? If not, how is optics a science conveying true knowledge about the physical world? Considered in Aristotelian terms, these are questions about the ontological status of quantity in a mixed science.

To answer these questions, we can turn to Philoponus' commentary on *Physics* II.2, one of the texts where Aristotle discussed the subject matters of mathematics, physics, and the mixed sciences such as optics. What we shall find is that Philoponus reworked Aristotle's thinking on the separability of mathematical objects so as to allow for the separability in thought of the "three-dimensional." This was an hypostasized version of Aristotle's bare continuity. Philoponus asserted that the forms of physical things are separable in thought from the three-dimensional, in a way that accords with matter-form distinctions present in the order of nature. This belief about the separability of physical forms from three-dimensionality is the methodological precedent to his separation of the directional component of visual rays from the mathematical lines of the visual cone. For separating in thought the subject matter for mathematical demonstration makes possible the consideration of the purely mathematical characteristics of rays of any kind. Then, the geometrical properties of the visual cone and the angles it makes at an object or mirror may be considered without regard to the direction of the rays. Once this is accomplished, it follows that rival physical theories may be evaluated in physical terms alone, as Philoponus urged in his commentary on 418b9.

The three-dimensionality Philoponus believed to be separable is not an undifferentiated extension but always a particular three-dimensional quantity. Philoponus believed that all physical forms are generated in a matter which is a mixture of elements specific both qualitatively and quantitatively to one particular form. Because of this, there is an indivisible and least quantity of any thing (*In Phys.*, 97.23). This three-dimensionality, also called by Philoponus secondary substratum (*deuteron hupokeimenon*) and body without qualities (*apoion sôma*), is part of a systematic theory of the layers of matter characteristic of the order of nature, which is held in common by the commentaries on the *Categories* and *Physics*. In his commentary on the *Categories*, Philoponus said of quantity that it is second only to substance among all the kinds of existing things. This is due to the fact that three-dimensionality is preceded only by prime matter as the most basic in the hierarchy of matters:

> For prime matter, as has often been said, being incorporeal and without form and shape, first having been expanded, receives three dimensions and becomes three-dimensional, what Aristotle calls secondary substratum; then in this way, it receives qualities and produces elements, so that quality has the third place among existing things and the fourth, things in relation (*In Cat.*, 83.14-19).

Philoponus invoked these layers in the order of nature in his commentary on *Physics* II.2. He said there are many matters of the same thing, some nearer and others more distant. He gave as an example the matters for the human body:

> For example, the nearest matter of the human body is the dissimilar-parted things of the whole body, as he himself says, but the matter before these is the similar-parted things. For these underlie in the order of matter those serving as organs. And before the similar-parted things are the humors and before the humors the elements, and before the elements the three-dimensional, and before this, prime matter ... (*In Phys.*, 232.1-6).

Philoponus believed that, because three-dimensionality was prior to substance in the order of nature, it was also prior in existence (*hupostasis*). In his *Corollarium de Loco*, Philoponus combatted the objection that quantity, as an attribute of substance, is dependent on substance for its existence. He said we could as easily maintain that substance is dependent on quantity, because of its reliance on a particular quantity for its makeup (*In Phys.*, 578.5-24).[1] Philoponus made this argument in the *Corollarium* to deny that the apparent dependence of quantity on substance should prevent our recognizing its priority in existence.

Two aspects of Philoponus' three-dimensionality--its separability in thought and its priority in existence--must be of concern to us in understanding the physical basis for mathematical demonstration for him. Mueller has shown that there is precedent in Aristotle for the conceiving of bare continuity in the manner Philoponus described, through the progressive stripping away of the attributes of body.[2] However, there is little to provide a basis for regarding quantity as prior to substance. Aristotle explicitly rejected the ontological priority of quantity to substance in *Metaphysics* Z.3, 1029a13-15, in a passage where he described the stripping away of attributes of body. Aristotle's commitment to the priority of natural substance is axiomatic in his natural philosophy, and the nature of focal meaning runs explicitly contrary to such a move. Aristotle's criticism of Plato on the separability of forms underlies his commitment to the priority of natural substance in the *Physics*. Nevertheless, there is reason to think that Philoponus found the philosophical justification for hypostasizing quantity in the issue of separability (*chôristos*) in thought and existence that Aristotle considered in his criticism of Plato in *Physics* II.2.

II

In *Physics* II.2, 193b35-194a7, Aristotle criticized those who hypothesize Ideas, for assuming that physical things are separable from body in the same way mathematical objects are separable. He said that man, flesh, and bone must be treated like 'snub nose' and not like 'concave.' The snub-concave distinction has two elements. The first is that we cannot talk about snub noses and snub bowls as we can talk about red noses and red bowls. Snub is inseparable from nose. The second is that snub is found in no other body besides nose. In the set of things that are snub, there is only one member: nose. 'Red,' on the other hand, may apply to many different kinds of material things. In this respect, 'concave,' is like 'red.' Flesh, bone, and man are all like 'snub.' In the set of things that are Man, there is only one member: man, meaning individual men. Aristotle thought that Plato, in making Forms separable, had ignored both elements of the snub-concave distinction. He thought that Plato's mistake was born of treating physical things as if they are just like mathematical objects. In making this criticism, Aristotle treated mathematical objects as themselves abstracted from sensible things (194a1).[3]

In his initial exposition of *Physics* II.2, Philoponus echoed Aristotle's view that the forms of physical things cannot exist apart from the bodies in which they reside, but he expounded the problem in terms of the kind of separability that characterizes different objects of thought. Like Aristotle, he distinguished 1) objects that can exist separately and independently (e.g., substance) from 2) objects that must exist as a part of some other thing that is separate and independent (e.g., attributes of substance). Some things that exist in this second way nevertheless are separable in thought (2a), while others are not separable in thought (2b). The forms of physical things belong to class 2b) for Aristotle, and Philoponus agreed with Aristotle on this point. Philoponus said it is impossible to conceive (*epinoêsai*) flesh or bone without the mixture of underlying elements that is the matter for each:

> If I tried to define flesh, for instance, I might say it is a warm, moist, bloody, soft body, having such and such actualities. But as soon as I have said this, I have included the underlying matter (*In Phys.*, 224.28-31).

It is not even possible to define a simple body, like water, without reference to matter. We conceive it as moist, fluid, and cool, but this is a description of water as a compound (*suntheton*) and necessarily involves the combination of form with its material base. Philoponus agreed with Aristotle that this is an argument against ideal Forms (225.4-5). There is no Man apart from the corporeal man. Philoponus recognized that the mathematician legitimately separates shapes and their attributes from physical objects, by disregarding the matter in which they reside (*In Phys.*, 224.17). Mathematical objects are separable from physical objects in thought, although they do not possess a separate existence. Physical forms are not even separable in thought, however, since it is impossible to disregard their matter in defining them. Platonists made the double mistake of separating in thought forms which are not separable in this way, and then granting independent existence to them as well (224.8-11). Mathematical objects are separable in thought (*kat' epinoian*) but not in existence (*kath' hupostasin*). Physical objects, however, are separable in neither way (225.2-3).

After this initial interpretation of the text, however, Philoponus went on to explain the extent to which he agreed and disagreed with Aristotle. He began by specifying that this criticism of Aristotle's was directed against Plato.[4] He agreed with Aristotle under the following conditions:

> I say that if Aristotle charged him with this: separating in existence (*to kath'upostasin chôrein idiai*) the forms of physical things, the accusation is reasonable (for it is impossible that things having being in a substratum ever exist in themselves (*kath'auta hupostênai*)), but if he means that it is impossible to separate form from matter in reasoning[5] (*tôi logôi*) or in thought (*tôi epinoiai*), it seems to me that it is not reasonable. For even if substantial forms (*ousiôdê eidê*) are themselves difficult to imagine in themselves, nevertheless reason (*logos*) can separate these too (*In Phys.*, 225.5-11).

We could characterize Philoponus' opinion in the following way: Aristotle was correct in thinking that the compound which is a physical object cannot exist apart from matter. Accordingly, our conception of the compound must include matter. To this extent, the form is not even separable from matter in thought. Nevertheless, Philoponus insisted, these same forms are separable from matter in thought. If it is possible to distinguish form from matter for physical objects, then it must also be possible to designate this matter:

> We say form is different from matter, and that matter, while remaining one and the same, receives differences in turn, since the forms are different from it. And what do I say is matter? The secondary substratum, then, by itself--and this is the three-dimensional--[which] is different from flesh and bone and all the remaining forms, being in itself unqualified (*In Phys.*, 225.11-16).

This substratum is, of course, the three-dimensionality which Philoponus believed to be prior to substance in the order of nature. He based his claim for some manner of separability of physical forms on the separability in thought of this body without qualities.

To separate this three-dimensionality, which is the particular quantity underlying and belonging to a body is not the same as separating mathematical objects. In the commentary on the *Physics*, a mathematical object is also a specific quantity but is considered apart from the object in which it resides. In the few references Philoponus made to the process of abstraction, he interpreted it as the separation of traits that could exist in a variety of bodies (*In de An.*, 62.24-27). Aristotle reserved for mathematical objects a particular substratum: continuity or multitude. Furthermore, even the substratum of geometrical objects he treated as abstractable.[6] In contrast, Philoponus defined mathematical objects without reference to a substratum at all. From this standpoint, mathematical objects may be regarded as themselves abstracted from sensible objects. For Philoponus, mathematical objects have a great deal in common with ordinary accidental attributes. He made this explicit in pointing out the difference between snubness and concavity:

> Just as if we define not snubness but simply concavity, we do not take note of matter at all, in this way also defining accidental attributes, we do not include the substratum in the definition (*In Phys.*, 226.21-23).

Accidental attributes, like mathematical traits, are not inextricably bound up with the substratum in which they reside, as are physical traits. As a result, they may be attributes of a variety of different kinds of bodies.

Philoponus distinguished, then, at least three different kinds of separability in thought, based on why the objects are separable in thought:

a) substances--because they are separable in existence.

b) mathematical objects and ordinary attributes--because they may belong to many different substrata.

c) forms of physical things--because they belong to a particular substratum that is itself separable in thought.

Based on this third manner of separability, we can see why Philoponus would claim that the physical hypotheses of optics can be examined apart from mathematical considerations--because the forms of physical things are separable from quantity, understood as a layer of matter. We can also see why the mathematical lines of geometrical optics have no direction--because the demonstrations of optics have to do with the lines in three-dimensionality which are unqualified in physical terms.

However, considerable problems attend this solution. For one thing, it is not obvious why being able to conceive the *matter* for substantial forms separately implies an ability to conceive those substantial forms separately as well. Presumably Philoponus believed that, in the case of a form-matter compound, to take away matter, even in thought, leaves form behind. He admitted that these forms may be difficult to conceive, and he did not elaborate on how to describe them. He was content to make the point that we can separate the matter in which these forms reside. Also, the question still remains of why manner of separability (c) is not at odds with Aristotle's principle that the forms of physical things must be conceived along with their matter. The answer to these questions turns, I believe, on the connection that Philoponus regularly assumed between the *conceivability* of three-dimensionality and its *existence* as a layer of matter in the order of nature. This connection itself relies on systematically developed beliefs of Philoponus about the relation between different ways of conceiving physical things and the way those things exist. One place to look for insight on this subject is Aristotle's *Posterior Analytics* II.1-10 and Philoponus' commentary on these chapters,[7] because it is here that Aristotle distinguished between the *ti esti* (what it is) and the *ei esti* (if it is) for a thing. In this context, Aristotle linked our ability to conceive what a thing is to our knowledge that it exists. He said, in chapter 8, that to have even something (*ti*) of the *ti esti* entails that we know the thing exists: "For it is impossible to know what [it] is, when we are ignorant if [it] is" (II.8, 93a20). Philoponus' commentary on these chapters may shed some light on his ideas about separability in thought and existence, because Philoponus involved separability in thought in his exegesis of these chapters.

III

In *Posterior Analytics* II.7-10, Aristotle considered the relation of definition to demonstration. This discussion is set in the larger context of determining if there can be demonstration of a definition. The second question is the concern of chapters three through six and takes the form of an inquiry into how *ti esti* is shown (*deiknutai*) (II.3, 90a36). In Aristotle's statement of his conclusions about the relation of definition and demonstration in chapter ten, he allowed for a sense of definition according to which the demonstration as a whole gives a definition. In the form of a demonstration, it is not called definition. But the demonstration may be formulated as a single sentence which includes the object's *dioti* (e.g., 'Thunder is a noise caused by the quenching of fire in clouds'). Aristotle called this a kind of demonstration of essence, differing from a true demonstration by the arrangement of its terms (II.10, 94a1-2). In this chapter, Aristotle allowed for another sense of definition, according to which 'nominal definition' (*logos onomatôdês*) (e.g., 'Thunder is a noise in clouds') may appear as the conclusion of a demonstration showing the *dioti*. Neither of these additional senses of definition is the indemonstrable account of essence that may appear as a premise of demonstration. There are thus three kinds of definition cited in chapter ten: 1) demonstration, or a definition differing from a demonstration only in the arrangement of its terms, 2) nominal definition, which is the conclusion of a demonstration, and 3) indemonstrable definition.[8]

At the end of *Posterior Analytics* II.10, Aristotle claimed he had made clear how demonstration and definition are related (94a14-19). Philoponus, in beginning his summary of these results, said that definition is of simples and demonstration of composites (*sunthetôn*). But there is also a kind of definition called the "composite definition" (*ho sunthetos horismos*) which is like a demonstration of *ti esti*, because it shows the *dia ti* (*In An. Post.*,

373.29-74.9). If we mean by *horismos* this composite definition, then there can be both definition and demonstration of the same thing. Philoponus then used eclipse to illustrate what he meant:

> For if you separate (*chôriseis*) eclipse from moon in thought (*têi epinoiai*), you show (*dêlôseis*) it [eclipse] through definition. But if you conceive (*theôreis*) it in the moon, you make evident through demonstration that [eclipse] is in it [moon] (*In An. Post.*, 375.21-24).

Some preliminary remarks can be made about Philoponus' use of the idea of separability in thought in this passage. First, it is clear that eclipse is not a simple thing and so there is no definition of it in the restricted sense of definition, i.e. indemonstrable definition. A definition of eclipse is a composite definition. As such, by Philoponus' own terminology, it includes a reference to cause ("Eclipse is deprivation of the moon's light due to the interposition of the earth between the sun and the moon.") Clearly, however, Philoponus believed this composite definition involves a separation of form from matter (eclipse from moon). The question then arises as to whether or not this separation corresponds to the kind of separability belonging to the forms of physical things. In spite of the similarity of language and subject, this passage does not make exactly the same distinction as the passage from Philoponus' commentary on *Physics* II.2, because in this passage from the *Posterior Analytics* commentary, the form of eclipse is not being separated from three-dimensionality, as a physical form from secondary substratum. Philoponus believed that the definition of eclipse applies not primarily to our moon and sun but to any bodies so configured as to share the same form. The form of eclipse is itself a function of the relation of shapes belonging to the celestial bodies involved. Thus, Philoponus would subsume the form of eclipse under the mathematical demonstration applicable to the bodies involved.[9] Consequently, with eclipse, we are dealing with a separation of mathematical characteristics from physical things, rather than a separation of physical forms from *apoion sôma*. Nevertheless, the similarity of language and the fact that both involve a separation of form from matter suggests there may be a relation between the kinds of separability in thought delineated here and in the *Physics* commentary. To explore this issue more thoroughly, let us consider a passage from Philoponus'commentary on *Posterior Analytics* II.2, where he treated the relation of definition to matter.

In chapters one and two of Book II, Aristotle said there are four things that investigation may seek: *ei esti*, *ti esti*, *hoti esti*, and *dioti*. Traditional interpretation has *ei esti* and *ti esti* applying to questions about the existence and whatness of a given object. *Hoti esti* and *dioti* apply to questions about whether an object possesses a given attribute and why it possesses that attribute.[10] Philoponus shared this traditional interpretation, referring to the *ei esti* and *ti esti* as "simples" and the *hoti esti* and *dia ti* as "composites."

It has always been a problem of interpretation, however, that in chapter two, Aristotle treated *ei esti* and *hoti esti* as if both apply to the possession of an attribute and *ti esti* and *dioti* as if both apply to the cause of possessing an attribute. In his commentary on the beginning of chapter two, Philoponus noted Aristotle's tendency to assimilate *ei esti* to *hoti esti*, and *ti esti* to *dia ti*, and gave reasons for Aristotle's doing so. For the assimilation of *ti esti* to *dia ti*, he gave the following reason, using eclipse as an example:

> The *ti esti* and the *dia ti* he makes the same insofar as the *ti esti* seeks the definition of the thing, when eclipse itself is taken (*lêphtheisês*) in itself, while the *dia ti* takes (*lambanei*) this definition as a middle term in a demonstration of why the moon is eclipsed (*In An. Post.*, 337.24-28).

Philoponus here treated the *ti esti* as if it refers to a composite thing and not a simple. Aristotle himself did this, Philoponus said, because he knew there could be no middle term involved in knowing the *ti esti* of a simple existent. For this reason, Aristotle took as his examples attributes of things (*In An. Post.*, 339.17-20). In a later passage, Philoponus said the reason for treating *ti esti* like *dia ti* is that Aristotle's concern in these chapters is with the question of how and in what cases demonstration proves a definition. For this reason, Aristotle was concerned only with definitions of traits present in a substratum, not with simple existents that are defined by indemonstrable definitions (*In An. Post.*, 369.5-9).

The most important part of Philoponus' discussion of the relation of definition to matter, in his commentary on *Posterior Analytics* II.2, is his contrast between the ways we think about simples and composites:

> And we say in the case of simple substances--that is, of intelligibles that are conceived (*theôroumenôn*) without matter, as are mind and soul--in the case of these, the form, i.e. the definition,

> can not possibly be something different from the definable (*to horiston*). <For example,> the definition of mind is 'potentiality of soul to be receptive of divine and intelligible things.' *This* definition, then, and mind are the same. And if they are one and the same thing, how is it possible to take a middle term between them? But in the case of composites that are conceived along with matter, for example man, even if the definition of [man]--mortal, rational animal--is taken in combination with (*suntetaktai*) man and conceived in him, nevertheless it is possible to take it separately (*kechôrismenôs touton labein*), and when it is taken separately, it seems [to be] something else (*heteros tis*) besides man. And for this reason in such cases, it is possible to take a middle term as cause. (*In An. Post.*, 339.6-16).

There is only one way of conceiving simple things. A definition is the single conception possible of a thing existing apart from matter. But for composites, besides the definition, there is the conception of the definition in some matter. For a composite, only the definition combined with a matter captures what the thing itself is in the way the mere definition of a simple does. There is thus an asymmetry in the function of definition for simples and composites. To understand this asymmetry better, it is necessary to take note of Philoponus' choice of examples to contrast to composites.

Philoponus' examples in this passage suggest he meant a contrast between simples, as substances existing without a substratum, and composites, as things that have their existence in a substratum. However, Philoponus characterized composite definition in terms of its including a *dia ti*. While this implies that the composite involves something present in a substratum, it does not suggest the composite's contrary is something that does not involve matter, but rather something whose cause is within, not outside itself. Early in the commentary on *Posterior Analytics* II, Philoponus made clear that he meant by *ta hapla* things whose causes are their own natures (*In An. Post.*, 336.12-23). He used Aristotle's examples of god, centaur, and moon, as examples of simples. In this context, he said that the investigation of simples concerns some one single term (336.21-22). He reiterated this point at the beginning of this passage contrasting mind and soul to composites, saying what is investigated in the case of simples is a single term (339.5-6).

The problem is that man would seem to qualify as both a simple and a composite on Philoponus' rendering of the terms. Man has all the causes appropriate to a generated being. But as a separable substance, man is the cause of himself and may be taken as a single term. Since Philoponus placed man among the composites here, the conclusion is inevitable that in this passage he *was* drawing on a contrast between substances capable of existing without matter and those that must have matter to exist. He added this contrast to the basic contrast of a thing having a cause within or outside itself.

His strategy in this regard is instructive, however, since his contrast between soul and man follows closely Aristotle's own contrast of these two in *Metaphysics* H.3. In this chapter, Aristotle said it is sometimes not clear to us whether a name signifies a composite or a form (1043a29-31). This need not concern us in the case of perceptible substance. In the case of perceptibles,

> ... the essence belongs to the form and actuality. For soul and being soul are the same, but [being] man and man are not the same, unless soul is also called man. Hence, <being man is the same> as one, but not the same as the other (1043a1-4).

This passage presents considerable problems of interpretation.[11] Most important for our purposes, however, is that Aristotle suggested an asymmetry in the relation of the composites--soul and body, on the one hand, and man, on the other--to their respective essences. Soul and being soul are the same, because form understood as essence is what makes the composite what it is. Aristotle seemed to be suggesting, by his contrast to the case of man, that soul itself has an essence, the same essence as belongs to the composite of soul and body. In contrast to soul and being soul, man and being man are not the same. This could be the case only if man is a composite in a different way from the composite of soul and body.

What Aristotle distinguished here were two different ways of conceiving a composite universally. Both rely on form to define the composite and both are intended to address the concreteness of form as identifying and determining the composite. But one (the relation of soul to its matter) does not identify matter as distinguishable from the actuality which its form gives it. The second (the relation of being man to man) does distinguish matter as different. The second is closer to our ordinary conception of the composite, but only the first addresses why the composite has the character it does, because the first formulation involves essence more directly.

This interpretation of *Metaphysics* H.3 is consistent with Philoponus' contrast of soul and man. Philoponus agreed that soul and being soul are the same, when he said the definition is not different from the definable for soul. He also associated the definition of man with form, and the definition taken in combination with matter with the composite. By using Aristotle's account of essence from *Metaphysics* H.3, he has distinguished two ways of conceiving a composite, the one an account of form alone and the other a formulation involving matter. Before developing the contrast between these two formulations more carefully, let us return to the passage from the *Physics* commentary where Philoponus delineated the kind of separability applicable to the forms of physical things.

There, Philoponus said that while the primary or proper conception of physical things required the inclusion of matter (if that conception is to avoid the mistakes of Plato), nevertheless, forms are separable from matter understood as three-dimensionality. Here, in the *Posterior Analytics* commentary, that primary or proper conception of physical things would appear to coincide with taking the definition of man "in combination with matter." Philoponus said, in this passage from the *Posterior Analytics* commentary, that this is different from another legitimate way of conceiving man, which is taking the definition apart from matter. This second way of conceiving man is like separating physical forms from three-dimensionality and also like separating eclipse from moon. Previously, the difference in what constitutes matter in these last two examples was noted to caution against identifying the separability treated in the two commentaries. Now, however, there seems to be reason to treat them similarly based on Philoponus' distinction between definition and definition taken in combination with matter. A recent interpretation of Aristotle's *Metaphysics* Z can broaden the basis on which the two commentaries are regarded as treating the same distinction.

In *Metaphysics* Z.10, Aristotle distinguished 'snub' from 'circle' on the basis of whether or not the definition must include reference to matter understood as the parts of the thing. Flesh is a part of snubness, although it is not a part of concavity. Bronze also is part of statue, when statue is considered as a compound (*sunolon*), but not when statue is taken as form (1035a3-9). Aristotle distinguished between snubness and statue here, suggesting the definition of snub would have to include matter, while that of statue would not. However, Aristotle contrasted both 'statue' and 'snub nose' to 'circle' based on the necessity of both the form of statue and snubness existing in compound with some material. Irwin thus noted that, based on his example of statue, Aristotle claimed "both that matter is part of a living organism's form and essence and that it is not."[12]

Aristotle's solution, Irwin believed, is to distinguish between proximate and remote matter. Both statue and snub "essentially include some definite type of proximate matter, but do not essentially include any definite type of remote matter."[13] Irwin went on to associate with these two types of matter two corresponding types of compound. The *formal* compound is a compound of form and proximate matter, while the *material* compound is a compound of form and remote matter. This latter distinction rationalizes Aristotle's apparently inconsistent claims about definition and essence. Definition of the *formal* compound, in the case of statue and snub, must include matter, but definition of the *material* compound need not. So there can be essence, conceived solely as form, which does not include matter, even for a object that must exist in some matter. There can be such an essence, because form is separable from remote matter.

Philoponus referred to the seventh book of Aristotle's *Metaphysics* in his introduction to his commentary on *Posterior Analytics* II (*In An. Post.*, 334.18). He probably drew on *Metaphysics* Z for the doctrine found in the passages from the two commentaries treated here. Certainly, there is much in Irwin's interpretation that Philoponus would endorse. Philoponus made the forms of physical things separable from remote matter, *apoion sôma*, but not from proximate matter. And his distinction between definition and definition taken in combination with matter serves to distinguish the two kinds of conception appropriate to the two kinds of compounds implied by this new separability. Definition is essence without matter, like statue without matter, or being man without remote matter, or the forms of physical things without three-dimensionality. Definition taken in combination with matter is the conception of the composite which is inseparable from matter. Philoponus was presenting two different ways of conceiving the universal. One way characterizes form, definition, or formal cause for a composite. The other way characterizes the composite itself. As Philoponus said, when we look at definition from the standpoint of the second conception, the composite, we notice that definition is actually "something else besides man," that is, something else besides the separable substance, man, taken universally as a compound of matter and form. Accordingly, being man is different from man. For substances that can be conceived without matter, on the other hand, definition and the thing itself are not different. On this latter point, Philoponus was clearly understanding soul as immaterial substance having an essence.

There is, thus, this asymmetry in the function of definition for simples and composites. For simples, definition and the definable are the same; for composites, definition does not capture what the thing is but only the definition taken in combination with matter captures this. Because there is a difference between the two conceptions for man or any other physical composite, there can be a middle term involved in knowing them. Philoponus built on these two conceptions slightly different meanings for *ti esti* and *dia ti*, in the case of a composite. He said that for understanding a composite thing like eclipse, we make use of both, but we use them differently:

> If we take the eclipse itself by virtue of itself (*autên kath'autên*), we seek what is the cause of eclipse, and we say 'privation of the moon's light due to screening by the earth.' But if we seek if eclipse belongs to (*huparchei*) the moon--that is, why it exists--we take the *ti esti* as a middle term--that is, the occurrence of privation of the moon's light due to screening by the earth. (*In An. Post.*, 339.25-29)

Definition is different from the conception of the definition in a substratum, but definition functions as a middle term in a demonstration proving a trait exists in a given subject. Note that the context for this discussion of *Posterior Analytics* II.2 is still why *ti esti* can be assimilated to *dia ti* in the case of attributes of body. However, in this context, Philoponus separated *ti esti* and *dia ti* in this passage, associating *dia ti* with determining the existence of an attribute in a subject. *Ti esti* does not involve existence while *dia ti* necessarily implicates existence. This is consistent with Aristotle's frequent insistence in the *Posterior Analytics* that what is proven in demonstration is the existence rather than the whatness of a trait belonging to a subject.[14] However, this point has importance in the context of Philoponus' separation of the forms of physical things in the commentary on *Physics* II.2.

In a passage immediately following Philoponus' comparison of concavity and accidental attributes in the commentary on the *Physics* II.2, Philoponus considered the mixed sciences--optics, astronomy, and harmonics--in order to show that even they cannot conceive their subject matters as simply mathematical objects apart from matter. Optics treats the properties of the visual cone in a purely geometrical way, but since it takes the lines and their properties as existing actually (*tauta lambanei energeiai huphestêkota*), it does not take them apart from motion and matter (226.27-227.14). Philoponus continued:

> Then, if those things that it is possible to separate in thought (*tôi logôi*) are not even separable in thought, whenever they are taken as existing actually, how much more is it unreasonable I suppose to separate things that are not even separable in thought (*têi epinoiai*). And these are substantial forms (*In Phys.*, 227.11-14).

Lines as mathematical objects are separable from matter in thought. They are not separable from matter in thought as rays with a physical character. The forms of physical things are like the rays possessing a nature. Without the insight gained from his commentary on the *Posterior Analytics*, it might seem that Philoponus did not leave much room here for the mode of separability in thought (c) that belong to the separation of three-dimensionality from physical forms. There is either the line geometry treats or the ray having a nature. The clue to his accommodation of this new mode of separability is that Philoponus made the difference between whether or not the lines of optics are separable depend on whether they are "taken" as existing in actuality (*energeiai huphestêkota*). *Energeia* and *huphestêkota* used together here denote something that exists and whose nature is fully actualized so that nothing more is required to complete its form. This mode of 'taking' is in accordance with the conception of the universal which includes matter with definition, in the commentary on *Posterior Analytics* II.2. Seen in relation to Aristotle's criticism of Plato, this mode of 'taking' is the way the forms of physical things are capable of existing--in combination with matter. However, there is another legitimate way of taking these forms, one that does not take them in the way they are capable of existing.[15] This is the separation of forms in accordance with definition that gives the essence of a composite like man. This conception--a formulation of the *ti esti*--does not implicate existence in the same way the other conception of the universal does.

Philoponus' language in the passage about optics suggests he believed it was possible to conceive a thing in the way it is capable of existing and to recognize this conception, distinguishing it from other ways of conceiving the same thing. His discussion in the commentary on *Posterior Analytics* II.2 adds weight to this impression. I have suggested that the reason for Philoponus' hypostasizing of three-dimensionality had to do with the relation between conceivability and existence in his thinking. It can be seen now, in conclusion, that the separability of three-dimensionality in thought is consonant with differences Philoponus saw in how essence may be conceived.

Certainly, Philoponus' discussion of separability in thought in the two commentaries makes more understandable his view of the mathematical science of optics. In nature, the lines of geometrical optics are inseparable from the light rays and rays of actualities of colors present in the air. This does not, however, preclude our separating the physical nature of the rays from their role in mathematical demonstration. In demonstration in optics, the rays need have no direction. In the evaluation of physical arguments about the nature of light and vision, geometrical optics need play no decisive role.

Notes

1. On the relation of Philoponus' idea of three-dimensionality to Simplicius' commentaries and to Philoponus' own later views in the *Contra Proclum*, see Richard Sorabji, *Matter, Space, and Motion* (Ithaca, N.Y.: Cornell University Press, 1988), ch.1 and 2.

2. Ian Mueller, "Aristotle on Geometrical Objects," *Archiv für Geschichte der Philosophie* 52, no.2 (1970):156-71; reprinted in *Articles on Aristotle*, ed. by J. Barnes, M. Schofield, and R. Sorabji (London: Duckworth, 1979), vol.3, 96-107. See Mueller's recent follow-up to this article, "Aristotle's Doctrine of Abstraction in the Commentators," in *Aristotle Transformed*, ed. by Richard Sorabji (Ithaca, N.Y.: Cornell University Press, 1990), 463-480.

3. Other passages relevant to Aristotle's discussion of the snub-concave distinction in relation to separating the subject matter of mathematics and physics are *De Anima* I.1, 403b7-16, III.4, 429b10-22, III.7, 431b2-17; *Metaphysics* E.1, 1025b30-26a10, K.3, 1061a28-b4. Aristotle used two different senses of abstraction, one the abstraction of mathematical objects from physical things, the process Aristotle described in *Physics* II.2, the other the abstraction of sensible qualities away from extension, as in *Metaphysics* K.3, 1061a28-35. Mueller ("Geometrical Objects," 100 (*Articles*, vol.3)) pointed out these different sense of abstraction give rise to two different views of mathematical objects. Philoponus made use of both senses of abstraction, as will be shown in Chapter Seven.

4. Simplicius said instead that Aristotle's criticism was directed against the popular conception of Plato's theory of Ideas. According to this popular conception, there are Ideas for everything that exists in the Demiurge's mind just as they exist here below, along with matter. The relation between things and their ideas is one of sameness, rather than similarity between an image and its paradigm (*In Phys.* [Simpl.] 295.12-18). Simplicius believed that Aristotle thought the causes of things in the generable world preexist as Ideas in the Demiurge's mind (295.21-32). Like Philoponus, he thought that physical forms could not even be separated in thought (293.31-294.9). Nevertheless, these forms are similar to Ideas in the Demiurge's mind and their definitions are really the definitions of these Ideas. The definition of snubness, for example, must be given in terms of concavity (294.12-18). There is no definition of 'nose' at the level of the Ideas. Later, Simplicius

admitted that it is possible to separate physical forms in thought, as witnessed by the fact that we discuss the form of a thing apart from its composite substance. What we discuss is not the exemplar, however, but something merely similar to it (296.10-25). Philoponus made no reference in his commentary on this passage to the causes of physical things preexisting in the mind of the Demiurge. He linked the separability in thought of physical forms to the existence of corporeal three-dimensionality as the matter for these physical forms. Simplicius referred to three-dimensionality only in relation to the subject matter of mathematics and did not grant it existence prior to substance, as Philoponus did.

5. While 'in word' might seem a more natural translation of *tôi logôi* in this passage, other uses of *tôi logôi* in relation to *têi epinoiai* indicate he means by *tôi logôi* not simply spelling out that a given thing has a form and matter, as the translation 'in word' would suggest. He rather means something interchangeable with *têi epinoiai*. See *In. Phys.* 227.11-14. quoted below.

6. See, in particular, *Metaphysics* Z.7, 1029a16-19, 11.1036a30-b3. On whether the continuous should be treated as the genus of mathematical objects or kept as substratum in a different sense than genus, contrast Mueller, "Geometrical Objects," *Articles*, 104, and Robert Sokolowski, "Matter, Elements, and Substance in Aristotle," *Journal of the History of Philosophy* 8, no.3 (July 1970): 279. The implications of this difference are considerable for our understanding of how the matter for mathematicals is in the world. Although much of Philoponus' account of the three-dimensional is consistent with Mueller's outline of Aristotle's position, Philoponus would not go on to identify Aristotle's extension with a rationally apprehended space, as Mueller suggests (107), but rather with quantity as secondary substratum. We might describe Philoponus' position as being more physicalistic than rationalistic, concentrating on extension as the stuff for physical forms.

7. M. Wallies, the editor of Philoponus' *In Analytica Posteriora*, doubted the attribution of the commentary on Book II to Philoponus. See his preface in *Comm. in Ar. Gr.*, vol.13, part 3, v-vi. To avoid circumlocution, I have referred to the author of the commentary on Book II as Philoponus. The convergence of thinking between the commentaries on *Physics* II and *Posterior Analytics* II suggest that, if *In Analytica Posteriora* II is not by

Philoponus, it at least represents the point of view of the Alexandrian school.

8. On the kinds of definition listed in chapter ten, see W.D. Ross, *Aristotle's Prior and Posterior Analytics* (Oxford; Clarendon Press, 1949), 634-36. Some see four types of definition given in this chapter. See David Demoss and Daniel Devereux, "Essence, Existence, and Nominal Definition in Aristotle's *Posterior Analytics* II.8-10," *Phronesis* 33, no.2 (1988):137. Other treatments of the issues in these chapters are Robert Bolton, "Essentialism and Semantic Theory in Aristotle: *Posterior Analytics* II,7-10," *Philosophical Review* 85 (1976): 514-44; J.D. Ackrill, "Aristotle's Theory of Definition: Some Questions on *Posterior Analytics* II.8-10" in *Aristotle on Science: The* Posterior Analytics, ed. by Enrico Berti (Padova: Editrice Antenore, 1981): 359-384; Richard Sorabji, "Definitions: Why Necessary and In What Way?" also in the Berti volume.

9. See Chapter Seven for more on Philoponus' reasons for subsuming demonstration in a mixed science under the relevant mathematical demonstration.

10. For an account of the traditional interpretation, see Ross, *Analytics*, and Suzanne Mansion, *Le Jugement d'existence chez Aristote* (Paris: Declée, 1946), 163-65. For a dissent from the traditional interpretation, see Alphonso Gomez-Lobo, "The So-called Question of Existence in Aristotle, *An. Post.* 2.1-2," *Review of Metaphysics* 33 (1980): 71-89.

11. On the interpretation of this passage, contrast Ackrill, "Aristotle's definitions of *psuchê*," *Articles on Aristotle*, vol.4, 68, and Irwin, *Aristotle's First Principles* (Oxford: Clarendon Press, 1990), 243-44. Ackrill thought that, when Aristotle talked about 'what-it-is-to-be-E (essence),' he misleadingly compared a composite, for which there is a distinction of object and essence, to an essence, for which there is no such distinction. Irwin, on the other hand, believed Aristotle meant form and compound have the same essence, for a compound like soul and body. I have emphasized Irwin's interpretation because of its relevance to Philoponus' formulation in the passage under consideration. See also Joseph Owen's notes on *Metaphysics* H.3 in his treatment of essence in *The Doctrine of Being in Aristotle's Metaphysics*, 3rd ed. (Toronto: Pontifical Institute of Medieval Studies, 1978), 183, note 85, and 187, note 88.

12. *First Principles*, 240.

13. *First Principles*, 242.

14. *Posterior Analytics* II.7, 92b15-16; Philoponus' *In An. Post.* 360.15-19, 361.10-18.

15. Philoponus' discussion of separability and conceiving in the two commentaries utilizes similar terms. In these passages, *lambanein* means the taking of a term or a trait itself so that it becomes either a definition or part of a demonstration, either a form separated from matter or one taken with matter. *Theôrein* is used to describe the mental act of taking something in one of these two ways. In passages already cited, *theôrein* is linked to *chôrizein*. *Epinoêsai* is used in a way similar to *theôrein*. In addition to the passsages already cited, see for example the following: *In An. Post.*, 340.29-32, 341.1-6, 343.24-26, 361.11-15, 364.16-18, 369.5-12.

Aristotle used *lambanein* in the *Prior* and *Posterior Analytics* for positing premises for a syllogism or demonstration, or for the positing without proof of the meaning of terms or of their existence. Ackrill associates Aristotle's use of *labein* with a pre-demonstrational knowledge that a conclusion is true or a particular middle term is applicable "... not full-strength knowledge but only knowledge in a diluted sense." See his "Aristotle's Theory of Definition: Some Questions on *Posterior Analytics* II.8-10," 366, 378.

Chapter Seven

Mathematical Demonstration in a Mixed Science

I

There are two contrasting uses of the concept of abstraction in Aristotle, one that sees abstraction as the separation of mathematical objects from sensible objects, the other as the stripping away of sensible traits from a rationally conceived extension. Mueller has said that Ammonius and Philoponus adopted the first concept of abstraction, making mathematical objects traits separable from any matter at all. He believed this doctrine can be traced back to Alexander of Aphrodisias, though he does not believe it could be a correct account of what Aristotle himself thought about mathematical objects. Mueller sees Aristotle's abstractionist position as being of the second type, involving the abstraction of sensible properties from three-dimensional extension.[1]

Philoponus' treatment of separability in thought in the commentary on *Physics* II.2 suggests how these two kinds of abstraction may be compatible. Philoponus made sensible forms separable from their matter, three-dimensionality. But this separability itself makes possible our ability to define the subject matter of mathematics as something separable from any sensible matter at all. From this, it is a short step to separating mathematical traits as if they are accidental attributes. As a result, in the actual practice of demonstration in a mathematical science like optics, demonstration may be fully subsumed under the appropriate geometrical demonstration. Philoponus articulated such a relation between geometry and the mixed sciences, optics, astronomy, and harmonics, in his commentary on *Posterior*

Analytics I. His views on this subject are developed in detail in the context of his discussion of the way demonstration in astronomy is universal.

II

Philoponus followed Aristotle in holding that the principles of mathematics are indemonstrable in the context of mathematics. Philoponus thought that they require demonstration by a superior science, first philosophy. He defined the principles of geometry as the point, line, plane, and those derived from these. The principles of arithmetic, on the other hand, are the monad and numbers, both odd and even (*In An. Post.*, 314.14-19, 320.30). In general, geometry treats continuous quantity, while arithmetic treats discontinuous quantity. On this view, we could suppose that the subject matter of mathematics has an independent existence, since in the *Physics* commentary, Philoponus made continuous quantity identical with the three-dimensionality underlying bodies. However, in the commentary on the *Posterior Analytics*, he treated continuous quantity as the raw material for mathematical objects, rather than the stuff of the universe. Consequently, in this commentary, the separability in thought of mathematical objects reflects the conventional view of quantity as an attribute of substance. However, in general, Philoponus tended to regard mathematical objects as considerably less important than the unique quantity which is taken up as matter in any physical thing. This quantity is reflected, in turn, in the measured place which that body occupies, place being a function of three-dimensionality. It would appear that mathematical objects, in the *Posterior Analytics* commentary, are objects of imagination which are derived from the discrete quantities which have actual existence in the world.

Proclus believed that true mathematical objects are implanted in the human mind by the Divine Mind. These differ from mathematical objects ordinarily manipulated in the imagination. This latter sort exist in extension and are divisible. But the student must strive to understand mathematical substance without the aid of these inferior representations. Proclus thought that mathematical objects hold a middle position between perceptible and intelligible objects because, while they are immaterial like the highest realities, they are subject to extension and division, and are not self-subsistent, like Plato's Forms. As a result, they are the province of the intermediate of the three rational faculties of the soul: opinion (*doxa*), reason (*dianoia*), and intelligence (*nous*).[3] It does not appear that Philoponus shared Proclus' reservations about extension and divisibility. It is precisely continuity, both extended and divisible, that he made the subject matter of mathematics. Like Proclus, he made mathematical objects the province of reason, but his explanation for doing so was different. According to Philoponus, mathematical objects have an immaterial being, but they are inseparable from perceptible objects. Because they do not exist apart from these objects, they are the objects of a faculty intermediate between intelligence and opinion (*In de An.*, 1-3.11). The explanation he gave for assigning mathematical objects to reason, that they are not completely separable from matter, implies that he regarded this continuity, which manifests itself in the mathematical objects of the imagination, as anchored in the world of physical reality. Philoponus never went so far as to identify the subject matter of mathematics with three-dimensionality. There appears to be a point by point correspondence between them. However, quantities have place while mathematical objects do not. The correspondence between the subject matter of mathematics and three-dimensionality is reinforced by the agreement between Philoponus' remarks in the commentaries on the *Physics* and *Posterior Analytics* about the subject matters of optics, astronomy, and harmonics.

Throughout the commentary on the *Posterior Analytics*, Philoponus reiterated the point that while different branches of knowledge may utilize the same common notions (axioms), their theses are different. In any syllogism, an axiom is the major premise and a thesis, defining an object or predicating something of a substratum, is the minor premise. There are sciences which make use of the same minor premises. This occurs when one science is a more general case of a subsidiary science. Geometry and optics are related in this way. Geometry can take the propositions of optics as its own, because the two sciences treat common substrata, i.e., mathematical objects (*In An. Post.*, 100.11). Optics does not contain all the propositions of geometry, since some are not applicable or useful to it. One example is the proposition that lines from the center of a circle to its periphery are equal. This proposition does not appear in geometrical optics (*In An. Post.*, 100.23-24). Geometry is both more general and more fundamental than optics, because it deals with mathematical objects taken in abstraction:

> For geometry treats lines simply and shapes simply, without considering (*episkeptomenê*) in which substratum they exist at some

> time, but conceiving (*theôrousa*) in abstraction lines themselves, angles themselves, and shapes themselves. However, optics conceives the lines in visual rays and the angles made particularly by them and the shapes (*In An. Post.*, 100.17-22).

As we have seen, in his commentary on *Physics* II.2, Philoponus' interpretation of the distinctions among physics, mathematics, and the mathematical sciences hinges upon the involvement of matter in explanation. Geometry, for instance, considers lines apart from physical bodies, purely as mathematical objects. Astronomy, harmonics, and optics take the objects of geometry and consider them not as mathematical objects, but as physical objects. These sciences do not separate mathematical objects but deal with them in matter (*In Phys.*, 227.20-27). But, even though the astronomer investigates the motions of particular heavenly bodies, the conclusions deduced from the more general mathematical discipline superior to it apply directly to astronomy (*In Phys.*, 220.30-34). This is the reason why astronomy qualifies as a demonstrative science. One could argue that astronomy does not carry the certainty of demonstration of universals, since it deals with particulars. Philoponus maintained, however, that although astronomy argues about particulars, it attributes traits to them not as particulars but as universals (*In Phys.*, 221.2-10). The reason why demonstration in mathematical science is universal is an issue of considerable importance in understanding the relation of physical theory to mathematical science for Philoponus. He derived astronomy's claim to universality from its corresponding mathematical demonstration. The basis on which he assimilated astronomy to mathematics is presented in his commentary on *Posterior Analytics* I.8.

In *Posterior Analytics* I.8, Aristotle claimed that universal demonstration is only possible for eternal objects (75b21-23). He said that demonstration of corruptible things may not always hold, since those things exist at some times and not at others (75b24-30). Demonstration of things that happen often, like eclipses, holds always (*aei eisin*) insofar as it is demonstration of a particular thing, *qua* that thing. However, insofar as these objects are not eternal, demonstration concerning them is partial (75b32-35). Thus, demonstration of recurrent phenomena appears to fall between universal and particular demonstration. The idea seems to be that whenever an eclipse occurs, it is always due to the same causes, because the original demonstration was based on *per se* attributes.

The problem with this view is that insofar as demonstration of recurrent phenomena treats *per se* attributes, it does not differ from particular demonstration. Even for a potentiality that is actualized only once, as long as demonstration about it is based on *per se* attributes, that demonstration will always hold in the same sense as demonstration about recurrent phenomena. The key to understanding universality in this sense is that the demonstration applies always to the same individual or individuals: a singular event or eclipses of the same objects. Aristotle believed, however, that demonstration of recurrent events ought to have a stronger claim to universality than the demonstration of unique events, simply because the events themselves are closer to eternality, involving as they do the same individuals in a repeating configuration. Thus, although Aristotle ascribed the universality of demonstration of this sort to the fact that it treats *per se* attributes, the only thing that distinguishes it from the *per se* demonstration of particular events is the recurrent character of the phenomena themselves.[4]

Philoponus, in his commentary on *Posterior Analytics* I.8 as in his commentary on *Physics* II.2, treated universality of demonstration in terms of its application to many other particulars of the same form, rather than in terms of the eternal or corruptible nature of the objects. In the case of eclipses, which are corruptible phenomena, he said that demonstration is not of any particular eclipse occurring at any particular time, but of the common form of eclipses. Even if there were ten thousand moons, they would be eclipsed due to the interposition of the earth between sun and moon (*In An. Post.*, 110.21-26). The fact that Aristotle thought universal demonstration is only possible for eternal things reflects his conviction that demonstration does not hold when the subject of demonstration does not exist. However, he immediately confused this point by saying that insofar as demonstration is of *per se* attributes, it always holds. Philoponus appears to have combined the two points. He said that demonstration in the case of eclipses is "according to the common form of all those eclipses, as many as fall under the same cause"(*In An. Post.*, 110.20). While Aristotle's point seems to be that the same sun and moon will produce an eclipse *at other times* for the same reason, Philoponus thought instead that other suns and moons just like those of our universe would submit to the same cause, because they are formally identical with our own sun and moon. Philoponus appears to have allied the kind of universality possessed by demonstration of eclipses to his own interpretation of demonstration, which is incidental (*hôsper kata sumbebêkos*) (75b24).

Philoponus said there can be no demonstration of particular statements like 'Socrates is an animal,' because Socrates is corruptible. But the demonstration is not purely incidental, because that of which demonstration properly is, itself exists in Socrates, namely the ensouled and perceptible, because this is in itself and primarily animal. The demonstration does not apply incidentally to Socrates, but rather 'at some time' (*pote*). That of which the demonstration is primarily must exist in the substratum, Socrates, and then the demonstration applies to Socrates (*In An. Post.*, 107.7-27). As Philoponus explained in his commentary on 75b26, such demonstration is partial, because not all the individuals to which it can apply exist all the time. Some have passed away, while others presently exist. This account bears some similarity to Aristotle's characterization of the limited form of eternality ascribed to demonstration of occasional occurrences, like eclipses. By events that happen often, Aristotle did not specify events always involving the numerically same individuals. This is implied in his use of eclipse as an example, since the sun and moon are unique individuals involved in a recurring pattern. If we consider recurrent events involving numerically different factors, then Philoponus' treatment of the problem in terms of different individuals existing at different times is justified. However, this approach is more relevant to the type of eternality possessed by biological species than by the phenomenon of eclipse. Philoponus did not treat the case where the same individuals are involved in the recurrent event. His remarks indicate, however, that he would have applied the same argument to this case as he did to individuals of the same species existing at different times (*In An. Post.*, 110.28-111.1).

It is important to note that the key to universal demonstration for Philoponus as for Aristotle, was that the demonstration concern *per se* attributes. The problem became one of particulars rather than perishables, as Aristotle originally formulated it, because of the priority accorded to a common form by Philoponus. This difference between Aristotle and Philoponus results in some important differences between their understandings of the relation of geometry to both optics and astronomy. This issue is treated in *Posterior Analytics* I.7 and 9.

In *Posterior Analytics* I.9, Aristotle said that knowledge is not accidental only when demonstration concerns the principles of a thing *qua* that thing. For the conclusion to constitute true knowledge, the middle term must be of the same genus as the extreme terms. There is a case where this does not happen, and knowledge is still of essential attributes. This occurs in demonstration in the mathematical sciences: optics, harmonics, and astronomy. In harmonics, the principles are arithmetical, but nevertheless, the demonstration is of *per se* attributes of the objects of harmonics. The reason is that the principles of arithmetic and harmonics have a common element (*to koinon*) (76a15). There are really two statements of the relation between arithmetic and harmonics in this passage. One the one hand, harmonics is subordinate to arithmetic, and on the other hand, the relation is symmetrical, based on a common element possessed by each. Barnes has offered an explanation of the first sort of relation.[5] Given that the proposition AaC belongs to subject matter y, if C is contained in B, and B belongs to a different subject matter x, then AaB belongs to x as well. We can imagine the relation between harmonics and arithmetic on this basis as being that of one set contained within the other. According to Barnes, this view, though tidy, trivializes the notion of a subordinate science. It can be criticized for being simplistic and ignoring completely the differences between harmonics and arithmetic. However, the meaning of this first sort of relation between mathematics and mathematical science becomes much clearer, upon considering the second sort of relation: symmetry based on a common element.

In *Posterior Analytics* I.7 and 9, Aristotle referred to demonstration based on a common factor three times. In the first case, he discussed demonstration based on traits which are not principles peculiar to the thing under consideration, like the fact that the straight line is the most beautiful of lines or that it is the opposite of the curved (75b17-20). For beauty and contrariness are not peculiar to line but exist in lines as traits held in common with other genera (75b20). In this case, the common factor means that the demonstration is accidental, not essential. Aristotle's second reference to demonstration by a common factor makes the same point. He said that Bryson's demonstration of squaring the circle relies on a principle that could apply to many other things besides circles or even mathematical objects (75b4-76a3). However, in his third reference, the common factor is the source of the identity of the principles of harmonics and arithmetic (76a10-15). For demonstration of essential attributes within one discipline, the common factor is in one genus, the same genus as the extreme terms. In the other two cases in *Posterior Analytics* I.7 and 9, where the common factor is found in many different genera, the demonstration is accidental, because it refers to traits that are not essential and thereby not unique. In the case of optics and geometry, the genera are discrete, but principles taken from each genus have something in common:

line\--------------> 'line' <-------------\visual ray

Geometry *Optics*

The same relation applies to the factor common to mathematics and any other of the mathematical sciences. The key difference between demonstration of this sort and *per se* demonstration is that the genera do not intersect nor does one genus contain the other. Rather, particular principles from each have in common something which is essential to the definition of each. This common element is itself indefinable, except in the context of a discipline. It is because this common element is *per se* to both principles, like 'line' to mathematical line and visual ray, that geometrical demonstrations can be used in optics. The common element is assimilated to its mathematical manifestation for the purposes of demonstration. It is in this more or less practical sense that optics is contained within geometry, or harmonics within arithmetic. However, their actual relation is better exemplified by the symmetrical relation of their principles based on a common element.

In his commentary on *Posterior Analytics* I.9, 76a15, Philoponus emphasized the relation of subordination between mathematical science and mathematics:

> For since he said that it is possible to demonstrate the principles of the subordinate science from the principles of the superior science, then also he adds the distinction, insofar as things demonstrated in this way differ from things demonstrated by means of their own principles, in order that no one should think that things demonstrated by means of the principles of the superior science are not proven, because of the fact that [they are not demonstrated] from their own principles. For this reason, he said that "the principles of these have the common element," i.e., the principles of subordinate sciences are common, like arithmetic and music, for demonstration in both cases is by common principles (*In An. Post.*, 118.26-33).

In this passage, Philoponus suggested that Aristotle introduced the common element only to insure that demonstration by principles of a superior science still constitutes proof in the inferior science. The importance of the relation of subordination is reinforced by Philoponus' remarks on the principles of optics, discussed above (*In Post. An.*, 100.11-24). Optics does not contain all the same propositions as geometry, but only some of them. In its turn, geometry cannot prove all its own principles. Its first principles are undefined in the context of mathematics but provided by first philosophy. First philosophy and geometry are not related to one another in the same way as geometry and optics. But the derivation of the first principles of geometry from a superior science reinforces the priority of the relation of subordination in Philoponus' treatment of the sciences. However, the subordination of optics to geometry has a formal significance it lacks in Aristotle, in that the mathematical demonstration defines the *per se* attributes of the science subordinate to it. This aspect of the subordination of mathematical science to mathematics plays a role in Philoponus' commentary on *Physics* II.2. He used astronomy's similarity to mathematics to explain how demonstration in astronomy is universal, even though it argues about particulars:

> Then, just as in arguing these cases involving particulars, the astronomer, in examining one moon and one sun, argues as if they pertain not to particulars but to universals, in this way, if he examines the solar sphere or its motion, he examines them not as they belong to the solar substance, but he examines them simply as traits subsisting in such a moving sphere, separating in thought the shapes and movements from substance (*In Phys.*, 221.14-19).

The traits the astronomer treats are *per se* attributes. For he explains the traits of a sphere moving in a particular way. Any sphere moving in a similar way would be subject to the same explanation. This is the implication of demonstration according to a common form considered in relation to mathematical science.

The assimilation of astronomy to mathematics in this way makes possible the separation of the subject matters of astronomy and physics. In his discussion of astronomy, Philoponus lapsed into a comparison of those who legitimately separate the attributes of substance and those who try to separate the forms of physical objects and grant independent existence to them (*In Phys.*, 221.20-222.2). This is a preview of his comparison of mathematicians and Platonists in the commentary on *Physics* II.2, 193b35. But in this context, he appears to be talking about astronomers rather than mathematicians. The interchangeability of astronomers and mathematicians in this respect is an indication of the close association Philoponus saw between the issues of the separability of physical forms and the subject matter of mathematical science. Philoponus made the comparison between mathematicians and Platonists in the context of explaining how physical forms are separable from their substratum, three-dimensionality. The appearance of this comparison in his discussion of astronomy suggests that Philoponus regarded the subject matter of astronomy as being just as closely related to three-dimensionality as geometry itself. The implication is that the demonstrations of astronomy can be considered independently of the physical forms of the objects involved.

This relation of subordination between geometry and the mixed science, astronomy and optics, is a function of Philoponus' treating mathematical objects as abstractable traits which yet define a *per se* subject matter. Forms of demonstration in astronomy and optics are mathematical forms to which the individual physical objects of the world conform. The priority of quantity to substance underlies the mathematical as the source of universality for demonstration in mathematics and supports the separation of mathematical traits in imagination. The priority of quantity to substance thus insures that the separation of mathematical traits does not produce solely mental entities but rather concepts that accord with the modes of existing that belong to an ontology released from the priority of natural substance.

Notes

1. Mueller, "Aristotle's Doctrine of Abstraction in the Commentators," 464-470.

2. Proclus, *In primum Euclidis Elementorum librum Commentarii*, ed. by G. Friedlein (Leipzig: B.G. Teubner, 1873), 3.1-5.10, 53.5-55,23.

3. On this point, see Jonathan Barnes' remarks in *Aristotle's Posterior Analytics* (Oxford: Clarendon Press, 1975), 131.

4. Barnes, *Posterior Analytics*, 153.

Chapter Eight

Conclusion

This book has approached two separate issues connected with Philoponus' interpretation of Aristotle's theory of light, one centering on his theory for the propagation of light, the other on his view of the nature of mathematical science. Aristotle's analysis of these issues reflects his preoccupation with the description of perception and thought. In each case, his account of the psychological and biological aspects of these functions of the soul originated in response to logical problems of the language used to describe them. His interpretation of visual perception in *De Sensu* VI and *De Anima* II.5 and 7 involved the distinction between motion and true actuality, a distinction which reflects the behavior of certain kinds of verbs used to describe action. His analysis of mathematical objects originated in his criticism of Plato's treatment of inseparable properties as separable. The snub-concave distinction associated with this criticism became the basis for his division of the subject matters of the various sciences. The presumption is that logical categories are based on nature--but as much on the nature of the soul as the world of which the soul has knowledge.

Philoponus' interpretation of these issues was more physical and cosmological. The propagation of light occurs in a way similar to the diffusion of heat and the communication of force at impact. A similar principle is involved in the way the soul acts and the Demiurge creates. Philoponus' account of how mathematical objects are separable from physical objects involved a theory of layers of matter. This theory portrayed a structure of the cosmos prior to the structure apparent to ordinary experience.

It is possible to find similarities between Philoponus' natural philosophy and later developments in scientific thinking in the West.[1] We could easily compare Bacon and Pecham's "multiplication of species" to Philoponus' theory for the propagation of light.[2] The similarity between the two theories of light should not be surprising, given the resemblance which has already been noted between Philoponus' remarks on projectile motion and medieval impetus theory.[3] However, in contrast to Philoponus, Bacon and Pecham held that light required time, albeit an imperceptibly short time, for its propagation. On this point, Descartes more closely approximates Philoponus' view. He thought light is a mechanical phenomenon, the communication of impact through a series of perfectly rigid incompressible bodies which then affected the eye.[4] This communication of impact must take place in an instant "in the same way as the action by which one end of a stick is pushed, passes to the other end at the same instant."[5] The principle behind this analogy is very similar to the one that underlies Philoponus' illustration of the simultaneous propagation of light:

> And, just as if someone moves the tip of a long and extended cord, the whole cord will move along with it without the passage of time, because of the continuity of parts, the first moving the next, this, it is necessary to think, is also what happens in the case of the actuality of light, due to the fact that all the successive bodies of the cosmos are in contact with one another (*In de An.*, 330.15-19).

Descartes' claim similarly rests on the supposition that all bodies of the universe are in contact with one another and that each translates to the next the effect it receives. However, for Descartes, the effect transmitted is simply impact. Light is an epiphenomenon of particles striking one another in a plenum. For Philoponus, light has a character apart from the material substratum that conveys it. Its nature is parallel to that of a physical body, however, in that it has a form. In these two respects, Philoponus' theory has more in common with the medieval theory of multiplication of species. Furthermore, in Descartes' theory, light is transmitted in the same way motion is translated between bodies. The body affected acquires just as much motion as the moving body gives up. The existence of a plenum assures that this communication of effect continues indefinitely. For Philoponus, the self-sustaining character of the propagation of light is due to the fact that the illuminating power transmitted to the medium from a light source serves as its own efficient as well as formal cause.

We might find a more legitimate basis for comparison between Philoponus and Descartes in their concepts of three-dimensional extension.[6] Each philosopher believed in a quantity of extension taken up in every substance. Furthermore, both maintained there is a correspondence between the subject matter of geometry and the three-dimensional basis of corporeal reality. However, for Philoponus, even three-dimensionality was a form, thought a simple one--mere quantity. To this extent, it conforms to the pattern of explanation used for light, where a variety of light's effects are attributed to aspects of its form. Even though Philoponus might share Descartes' tendency to find a simplest constituent of matter underlying all body, he did not give up the Aristotelian concept of cause--in this case, formal cause.

However, the changes in Philoponus' approach which we have already noted altered his use of the Aristotelian formal cause. Philoponus characterized the form of light in terms of light's observable effects of heating, illuminating, and rendering media transparent. Characterizing things in terms of constitutive attributes possessing perceptible manifestations is well-suited to providing a description of nature that supports the supposition of no extra entities. This is, at least in part, the task Aristotle set himself. The same program resembles a *petitio principii* when used in an explanation that treats perceptible attributes as parts of a thing. There is a clear tendency on Philoponus' part to regard attributes in this way. Hence, the properties of light, which are manifest by their effects on other things, assumed the character of incorporeal powers which could be transferred to the things they affect. Light was no longer simply an activated state of transparency, a trait of the medium. Philoponus believed it is an incorporeal actuality, which the transparent medium is capable of receiving and transmitting. Similarly, continuity was not, for Philoponus, a purely intelligible aspect of bodies but a substratum existing prior to substance.

However, at the core of this physical and cosmological language lay a causal analysis worthy of Aristotle's standard of logical perspicacity. Philoponus' commentary on *Physics* III.3 presents a neutral sense of the *energeia* concept which gives to the propagation of light and projectile motion a common minimal causal structure. This structure is one of the philosophical achievements of late antiquity and should remain a standard reference for the intensional meaning of causation. Furthermore, we have seen how this causal structure is not just a Neoplatonic gloss on Aristotle's concepts but a development of genuine problems of interpreting Aristotle.

It is possible to trace the direction imparted to Neoplatonic thinking by this cluster of Aristotelian problems. One such instance is the contrast between the activity of Nous and the work of a craftsman, discussed in Chapter Five. Citing Porphyry, Proclus explained the production of the universe by the Demiurge by contrasting it to the work of a craftsman who needs tools to implant the form in his material. However, if there were no impediment in the matter, the form would be produced in the substratum in no time whatsoever (*In Tim.*, I.395.10-22). This is precisely what happens in the case of Nous thinking itself. This contrast in the ways of conceiving production is one aspect of the mind-body problem in Neoplatonism. True actuality, as a vehicle for production, is creation by simple intentionality. The craftsman, on the other hand, must marshall a variety of materials, skills, and activities, in the service of that same intentionality.

Notes

1. This work has been begun by Richard Sorabji. See his *Philoponus*, ch.1, and his *Matter, Space, and Motion*, ch.2.

2. See Robert Bacon, *Opus maius*, ed. by J.H. Bridges (Oxford: Clarendon Press, 1897-1900), II.71-72, and John Pecham, *Perspectiva communis*, ed. by David Lindberg (Madison: University of Wisconsin Press, 1970), I.27. Relevant passages from various authors pertaining to the nature and speed of propagation of light are collected in Edward Grant's *Source Book in Medieval Science* (Cambridge, Mass.: Harvard University Press, 1974), 393-97.

3. See Marshall Clagett, *Science of Mechanics in the Middle Ages* (Madison: University of Wisconsin Press, 1959), 505-525. Establishing a material connection between Philoponus' theories and similar medieval theories is more difficult. It is not known if the relevant parts of Philoponus' commentaries on the *Physics* and *De Anima* were accessible. It is quite possible that medieval thinkers arrived at similar ideas independently or working with summaries of Neoplatonist philosophy. The availability of some ancient commentaries may have influenced the direction of medieval scientific thinking in a way that made the independent formulation of these theories likely.

4. On Descartes' theory of light, see A.I. Sabra, *Theories of Light from Descartes to Newton* (London: Oldbourne, 1967), 46-60.

5. Descartes, *Le Monde*, in *Ouevres de Descartes* XI, ed. by Charles Adam and Paul Tannery (Paris: L. Cerf, 1897-1910), 95. Sabra has traced the history of this analogy (*Descartes to Newton*, 54-55, n.34). It seems to have had nothing to do with the speed of light originally. The analogy is probably of Stoic origin. Alexander of Aphrodisias criticized the idea of *pneuma* proceeding out and shaping a visual cone from the air, through which it perceives by touch as if by means of a stick (*Mantissa*, 130.14-134.27). It is interesting to note that the analogy Philoponus used to illustrate the simultaneous propagation of light involved the propagation of effect through a non-rigid body. Perhaps, he used the example of a non-rigid body to avoid any association of his theory with the Stoic doctrine.

6. For a comparison of Philoponus' three-dimensionality to medieval and Renaissance ideas of space, see Edward Grant, "The Principle of the Impenetrability of Bodies in the History of Concepts of Separate Space from the Middle Ages to the Seventeenth Century," *Isis* 69, no. 249 (1978): 551-71. See also Sorabji, *Philoponus*, 21-22.

Appendix I

Analysis and Translation of Philoponus' Commentary on De Anima *II.7, 418b9*

(*In De An., 324.23-342.16*)

324.23-35	Exposition of 418b9
324.35-325.5	Plan for the commentary on 418b9 1. To refute that light is a body and that vision occurs by the emission of bodies from the eye 2. To set out the doctrine of Aristotle and examine the objections against it
325.6-326.37	That vision does not occur by the emission of bodies from the eye (includes 325.33-326.9, against the idea that an encounter takes place between visual rays and light rays; 326.9-15, against the idea that visual rays are lines; and 326.26-37, that nothing incorporeal is emitted from the eye)
326.38-329.2	That light is not a body (includes 328.13-21, against the idea of immaterial body)
329.3-330.27	Theory of light and vision based on Aristotle's doctrine of light 329.3-30 Air has the power for transmitting colors, being itself colorless 329.30-330.24 Light passes through the medium by a succession of effects

330.24-27 By means of this explanation, the forenamed problems are avoided: body will not pass through body, and it is not impossible that actuality comes to be without the passage of time

330.28-333.35 Presentation of objections
330.28-331.1 Reflection in mirrors and burning mirrors; friction and heating due to reflection of rays; that mountain ridges are heated less than the plain; that we do not see things behind us
331.1-10 Solution of these problems by the supposition that light and visual rays travel on straight lines and are reflected at equal angles:
331.10-19 That we do not see things behind us
331.19-33 Reflection in mirrors
331.33-332.22 Friction and warming
332.22-333.18 That midday is warmed more than dawn and dusk; that it is warmer in summer
333.18-35 Why Aristotle used the supposition of visual rays in the *Meteorology*

333.36-334.29 Problems raised against the reflection of incorporeal actuality: 1) how can something incorporeal make angles? 2) how can it be reflected apart from its own substratum? Solutions to these problems based on the natural ability of a reflecting body to actualize in return an actuality similar to that which it receives

334.30-341.9 Further objections considered
334.30-335.7 How do we know the distances between us and the things seen?
335.7-336.3 If the whole air is filled with images, why do we not see things without looking at them?
336.3-337.7 When some stain is floating in the eye, why do we not see that it is inside?
337.7-10 Why do we not see when we close our eyes?
337.10-338.5 Why do we not see the color of the iris?
338.5-339.16 Why do we perceive an affliction inside the eye to be outside?

339.17-30 What is the reason for loss of vision due to dilation of the pupil?
339.30-340.21 When one pupil is deflected from the undeflected straight line between the eyes, why do we see double?
340.21-341.9 How do we know distances and perceive magnitudes?

341.10-342.16 That darkness is the privation of light, and how this further demonstrates the incorporeal nature of light

418b9 "And light is the actuality of this, of the transparent *qua* transparent."

324.25 After he said what the transparent is, that it is the visible--but visible not by virtue of its own color but by virtue of that of another--he proceeded to a discussion about light, and he says that light is the actuality of the transparent insofar as it is transparent--that is, the form and perfection of it. For when the transparent exists potentially, the light that supervenes (*epigenomenon*), in giving it form and perfecting it, makes it / actually transparent. For the nature of the transparent is in air and water and such things; when there is no light, it is not actually transparent (for we would not see any color <through> it), but when light is present, it possesses its own form and perfection. For which reason, when light is present, we see through air or through water, but we do not see through earth, since evidently it does not have the nature of the transparent in itself potentially.

325.1 Some of those who have spoken about light, // have said it is a body. Others have said it is incorporeal. That light is not a body, and further that even vision does not occur by the emission of certain bodies from the eyes, it is necessary, after first refuting this very point, <showing> that it is impossible that either light or visual rays is corporeal,[1] then to set out the opinion of Aristotle concerning these / things and, in addition, to examine the difficulties raised against it.

If, then, seeing takes place by the emission of certain bodies, how is it possible that we perceive the heavens all at once, as soon as we look up? For how is it possible that a body moves so great a distance without the passage of time (*achronôs*)? For it is proven that the distance from the earth to the heavens is two signs of the zodiac. For the radius / is equal to the side of the inscribed hexagon, and the side of the hexagon is the distance of two signs of the zodiac.[2] And the circumference is three and one-seventh times the diameter; but the distance from the center is one-half of the diameter. Then, it is evident that, as the whole diameter is a little more than four signs of the zodiac, since the circumference is twelve, one-half of the diameter will be a little more than two signs of the zodiac. How is it possible, then, / for a body to move so great a distance without the passage of time? At any rate, the quickest of movements itself--I mean the movement of the sphere of fixed stars--moves in two hours the sixth part of the whole--that is, two signs of the zodiac.[3] How is it not unreasonable, then, to suppose that bodies emitted from the eyes move more quickly than even the sphere of fixed stars? Again, if we see by means of corporeal rays, since when we look up we see a / quarter of the heavens, it is absolutely necessary that the visual rays going through the air either pass through it, and in this way body would pass through body, which is impossible, or that they cut it. And if they cut it, we ought not to see the heavens or other visible objects in their continuity, if they do not pass through the whole air but cut it. It is evident, then, that the parts of the / visible objects behind the parts of the air that were not cut, would not be seen. And in this way, we would see no visible objects as continuous but as divided into parts.

Then (*epeita*), since by looking up, we see a quarter of the heavens, if seeing takes place by the emission of certain corporeoform rays, it is absolutely necessary to suppose the rays are such large bodies as to extend from us out to the / heavens and comprehend a quarter of it; wherefore, it is unreasonable to maintain that from a small organ, the eye, and not even from the whole eye but rather from the pupil, such large bodies are sent out as to comprehend a quarter of the whole. But against these objections, they say that a certain meeting (*apantêsis*) of visible objects and visual rays takes place. For certain lights are sent out from heavenly bodies which meet / visual rays in advance and permit the perception of themselves. But against this, we say that if vision does not contact the heavenly bodies themselves but the lights going out from them, how is it that astronomers know the magnitudes of stars? And their movements? And distances, both
326.1 conjunctions and oppositions, // and as many other things as they demonstrate concerning the heavenly bodies? For it is because they perceive the spheres themselves that they make their demonstrations concerning them.

Then, these lights going out from [the spheres] must be either corporeal or incorporeal. If, then, [they say] the lights are bodies, they will be refuted by the same arguments by which / we will next refute those who say that light is a body. But if incorporeal, what is the necessity for an emission of visual rays and a meeting in advance. For if actualities are entirely incorporeal, they will arrive at the eyes without a meeting even occurring; and thus they will say the same things as Aristotle concerning visual rays and light, as we shall see a little later. And in general, if seeing takes place by an emission / from the eyes, the emissions must be either bodies or lines, as the mathematicians seem to say. But if the visual rays going out are lines, it is evident that they will contact visible objects at points. But they say that seeing takes place by visual rays touching visible objects; but they touch [them] at points; clearly, they will also perceive visible objects at points. Clearly, they will not / perceive magnitudes, which is absurd. But if they are bodies, the part of the cornea surrounding the pupil must be pierced, which is false (for it is continuous, and as soon as it is pierced, the fluids flow out), or body will pass through body, which is itself also impossible, or visual rays must go out through certain pores. But that bodies are not composed of pores was shown / in *On Generation*.[4] Moreover, it would be absolutely necessary that there be certain particles and bodies in the tunics which separate the pores. Through those, then, it would be impossible for the visual rays to pass, so that we ought not to see the ground or anything else in its continuity, but [we ought] to see some parts and not others. For the parts of the visible object which the particles dividing the pores lie opposite, / these it would be impossible to see. Clearly, nothing would be seen in its continuity, which is contrary to manifest facts (*para tên enargeian*).

But neither is it possible that an incorporeal actuality goes out from the eyes. For it will go out either by aid of its own vehicle (*meta tou oikeiou ochêmatos*) or not. If by aid of its own vehicle, again bodies will go out, and the same arguments apply. But if it goes out separated from its own vehicle, it is absolutely necessary, I presume, that it come to be in the other / substratum. Since, then, the perceptive power is something living (*zôê tis*) (for what does not share in life could not share in perception), and life, in whatever it comes to be, makes that thing alive, clearly the perceptive power also, when it goes out of the eye and comes to be in the air or in water or, to put it simply, any transparent thing, shall make that [medium] alive, so that it will also be ensouled, which is ridiculous. Or if this does not happen, / there will be life in what is not living, which is impossible. Moreover, if the actualized optical power (*optikê dunamis energeiai*) goes out of the eye and comes to be in the air or in water or in some such thing, that [medium] will be something that sees and is capable of perception.

That nothing, then, either corporeal or incorporeal proceeds from the eyes, it is possible to show from many other arguments. Nevertheless, those that have been given are sufficient. And that
327.1 // light is not a body either, is evident in what follows. For if it were a body, again, how could the movement of a body take place all at once (*athroan*)? For as soon as the sun is over the horizon, suddenly without the passage of time, the whole hemisphere above the earth is filled with light. And if I carry a covered lamp into a room and then remove / the cover, all at once the whole room is illuminated. How, then, could a body be moved in this way without the passage of time?

Then, if light is a body, it is evident that, being a simple body, it will be moved with a simple motion. And simple motions are two: motion in a straight line and circular motion. If, then, light is moved in a straight line, either moving upward it will not move downward, or moving downward it will not move / upward. But as it is, it is seen to be moved upward and downward and to the right and left; but every [simple] body that moves straight is moved with either an upward motion or a downward motion only. But that [light] is not moved in a circle is evident from manifest facts. For if I introduce a lamp into a narrow rectangular room, from the entrance of the room immediately (*eutheôs*) the whole room is illuminated along straight lines. Holding it below, / the upper parts [of the room] are illuminated, and above, similarly, the lower parts [are illuminated]. But the specialists in optics also show clearly that light proceeds along straight lines and not in a circle. For if you pierce a plank or make a kind of slit in it, and on the other side of the plank you light a lamp at the slit, and on the other side place another plank, the light of the lamp going through the slit will appear (*genêsetai*) / on the other plank. If, then, you extend a straight line from the light appearing on the plank, it will pass through the slit to the lamp. So it is evident that the motion of light occurs along straight lines. So that if it does not move in straight lines as bodies do, either only upward or only downward, nor in a circle, and every body / has one of these forenamed motions, it seems that light is not a body. Again, if light is a body, it is absolutely necessary, if it passes through the whole air, that body be in body. And if it does not pass through the whole air but either cuts the air or passes through the pores of it, light must not be in every part of the air, but some parts of it are illuminated and some are in shadow, so that not / all the air will be illuminated. There will not even be air everywhere, but where there is light, there will be no air, and where there is air, there will be no light. And in this way, it will come about that we breathe light and air. And there will be times when we breathe light only and no longer also air, whenever we happen to be breathing in the illuminated region by itself. But these things are clearly absurd and contrary to manifest facts themselves. But also, when the wind blows, since it is a mass and continuous, it ought / either to completely darken the place where it falls, by displacing the body of light, or it should at least make the place darker, since it has displaced the body of light, or at least part of it, and taken their places. But as it is, this does not happen. We see // that even when the air is illuminated, it has the same characteristics (*hai autai tou aeros energeiai ginontai*). For we breathe and likewise we are cooled, although the nature of light is to warm, not to cool, even the air itself.

But in addition to these [problems], since those who maintain that light is a body say a certain sphere of light proceeds from / the sun, since the sun is spherical and the air also is spherical, it is absolutely necessary that the sphere of light proceeding toward us either push the sphere of air and, in this way, by occupying its place, illuminate the place around us, or else not push it but pass through it. If, then, it pushes the air and occupies its place, the illuminated hemisphere will be empty of air, / which is impossible. For as I said, all the activities of air go on (*panta ta aeros erga ginetai*), even when everything around us is illuminated. But if it does not push the air, neither will it produce illumination (*phôtisei*), since it does not reach us. But if it passes through the air, there will be a body penetrating another body, which is impossible. But if they say that, because a body is immaterial (*ôs aulon*), it can pass through body, what they say is utterly impossible. / For in this way the heavens could be in a grain of millet.

Then, if because of being immaterial, [light] can pass through body, it would pass not only through air but also through any other sort of body. For how likely is it that if a body, by being immaterial, can pass through a body it does not also pass through every body? So that light ought also to pass through solid bodies--that is, through earth and the / like. Why, then, when the sun is under the earth, do we not see above the earth, if light passes also through the earth?

Then, if light is a body and passes through the body of air, the whole ought to become denser and thicker. But, as it is, on the contrary, the air becomes finer when it has received light, which would not happen if it were true that light is a body. Just as, if water is diffused in earth and the place occupied by both does not become larger, / it is absolutely necessary that the body composed from both has become thicker, in this way also it is necessary that the air become thicker when light is present. But as it is, on the contrary, it becomes finer when illuminated. But if someone said, objecting to the account of light's motion, that certain rays are dependent (*êrtêmenai*) on the sun and illuminate by change of place of motion (*têi metabasei tês kinêseôs*) / without requiring motion from above (for they are always dependent on the sun) to reach the earth, he speaks most contrary to the observed facts (*ta phainomena*). For the sun and moon and fire and all other bodies illuminate in a similar way, but he will not be able to say this about fire or one of the others. And whenever, in the dark, I light a fire by the friction of sticks in a very large room, without the passage of time, / light fills the whole room. And when the air is thickened, the sun is not seen, but all the place around us is illuminated.

Then, I shall inquire about the rays of sunlight, whether they are corporeal or incorporeal, // and whether the rays are borne through the whole air or not through the whole, and again the same difficulties arise.

Moved by these and other arguments of a similar nature, we are compelled to say that neither light nor visual rays are bodies. What, then? How does seeing take place? / And what is the nature of light? We say, then, to begin with, following Aristotle on vision, that the transparent, air, has a power for transmitting (*diaporthmeutikên*) colors, while being itself colorless. For if it had some color of its own, this would hinder its transmission of other colors. For instance, if a glass is colorless, it transmits all colors. / But if it is tinted by some color, it neither transmits all colors (for it does not transmit the fainter ones) nor does it transmit accurately. The air, then, has in every part of itself the power of transmitting colors. Indeed, it is completely transparent throughout; for what lies here is seen by those in front and in back and, quite simply, by viewers in any part of the air. Certain actualities (*energeiai*), then, / from visible objects are produced without effect (*egginontai apathôs*) in the air [and] without coloring it, since it is not its nature to be colored. Whence different viewers see different colors through the same part of air. This would not happen, if the air were colored, for it cannot have contrary colors at the same time (*tois enantiois hama chrônnusthai*). For example, if O is the air, M is black, and L is white, and A sees the white object / and B the black object, the actualities of the white and black objects will pass through the air O. But if they are contraries and they pass through the air, it is evident that they do not color it. For it is impossible that the same part will receive contrary colors at the same time. What can be said about these two colors can also be said of more colors. For when there are more viewers and visible objects, / more actualities of colors come through the same air and meet (*sumpiptousai*) one another. But if the air were colored, this could not possibly happen.

The actualities of colors, then, traverse (*diabainousin*) the transparent and communicate (*apaggellousi*) their own substrata to vision--I mean, colors themselves. They act (*drôsai*) on perception itself in this way or that, but produce (*empoiousai*) no effect in the transparent. / And it is not necessary to inquire how the actuality of color, once it has left its own substratum, reaches vision. For it comes to be in the air as in a substratum. And in this way through [the air] it acts on vision, just as also the actuality of the carpenter passes through the axe to the wood and acts on the wood, giving it one shape or another, but in the axe it produces no such [effect] as it does in the wood. Just as, then, in these cases, so also in the case of light, we say a certain incorporeal actuality is sent out from the luminous body to

transparent things which are naturally inclined to receive it. And just as fire, even when it lies some distance away from us, warms us, not by its travelling (*phoitan*) to us, but by warming the air
330.1 next to it // which is naturally inclined to be acted upon by fire, and that [air], when it is warmed, becomes such as to be able also itself to warm the [parts] lying next to it, and this goes on (*ginetai*) until all [are affected], and in this way even things very distant from the fire are warmed--directly by the fire but proximately and secondarily (*prosechôs de kai kata deuteron logon*) / by [the parts of the air] near it--in just this way also, since the transparent bodies lie in succession (for they are spheres and touch one another, and all the air in the sublunary world is in succession to the last sphere, and water is in succession to it) the sun then acts proximately on the body next to and touching it and makes it actually transparent. And that body, once actually transparent, itself / has the power of acting on things nearby and making them [actually transparent], and in this way the illuminating power (*phôtistikê dunamis*) proceeds out to the limits, not by the actuality of the sun travelling without mediation to the limits, but by the things that are acted upon directly being able themselves to actualize the similar actuality (*tên homoian energein*)[5] in things capable of receiving it. If the actuality comes to be all at once (*athroa*) throughout the whole, this is not surprising, / seeing that the actuality is incorporeal. And just as, if someone moves the tip of a long and extended cord, the whole cord will move along with it without the passage of time, because of the continuity of parts, the first moving the next, this we must think is also what happens in the case of the actuality of light, because of all the successive bodies of the universe being in contact.

Light, then, is not a body, but the / actuality of the transparent *qua* transparent, which comes to be due to an efficient cause. The efficient cause in the primary and strictest sense is the sun, and secondarily as many as by sharing a certain emission (*aporroias tinos*) from the [sun] have become themselves capable of illuminating, just as have the moon, and the rest of the stars and fire. For God placed the primary source of light in the sun. In this way, then, we will escape all the puzzles / mentioned, and neither will body pass through body, nor is actuality unable to come to be all at once and without time (*achronon*), since it is incorporeal.

But they present puzzles to us concerning these hypotheses of light and visual rays, that if both light and visual rays are not corporeal, / how (*pothen*), then, does reflection in mirrors [take place], and also in burning mirrors? How is friction, and from this, warming, [produced] from the reflection of rays? And why are the mountain ridges heated less than the plain? And why, if the actualities of visible objects are in the air, do we not see also the things behind us? For if the air has the actualities of color in every part of itself--as is shown by the fact that / those looking from every direction see the same color (for example, if A is a color, and around it are many viewers, B, C, D, E, F, and G, all see it. It is evident, then, that this is due to the fact that the air has in every part of itself the actualities of colors)--and if this is true, why in the world do we not see also the things // behind us?

We will give one solution, then, to all these problems employing as hypotheses for the account of actualities those [hypotheses] which they use who say light and visual rays are bodies. Just as, then, they suppose (*hupotithentai*) that both visual rays (*opseis*) and [light] rays (*aktines*) are sent out along straight lines and are reflected / from smooth bodies at equal angles, in a similar way also, we suppose that both the actualities of colors, and furthermore even light, are sent out along straight lines and are reflected from smooth bodies at equal angles. For this reason, then, also images (*eidôla*) in mirrors appear, not by our visual rays being reflected at the visible objects, but by the actualities of [visible objects] / [coming] to us. For this reason, neither do we see things behind us, since the actualities are carried (*pherontai*) along straight lines to the eyes, and in no other way than along straight lines shall the actualities of visible objects be carried, unless we are face to face with them. For the actualities are always in the air without affecting it, but they do not act on us, except when the perceptible objects are on a straight line with the / eyes, as neither do visual rays, according to them, contact (*prosballousi*) the visible object, unless they are on straight lines with it. For this reason, then, even when we look sideways, we do not see the colors behind, although the actualities of colors are in the air, on account of the actuality not being carried along a straight line to the eye. It is in this way, then, that we say the reflections of light and of the actualities of visible objects too occur, / when the actualities are reflected from smooth bodies at equal angles. And for this reason, by the same account of visual rays, whenever a mirror happens to have a position with respect to us and the visible object such that the actualities falling on it and reflected at equal angles travel (*phoitêsai*) to the eye, we see even things that do not lie on a straight line with the eye, / e.g. things behind, or above, or below, or to the sides. And quite simply, whatever they say concerning visual rays, we say the same things in precisely similar terms concerning actualities and save the appearances. For what difference does it make whether straight lines come to the mirror proceeding from the eye, or are reflected from the mirror to the eye? If, then, these are held in common by those [thinkers] / and the doctrine of Aristotle--but the one in terms of actualities, and the other in terms of visual rays--and myriad impossibilities follow from the hypothesis of visual rays, one must rather choose the hypothesis of Aristotle which both preserves the appearances and avoids the impossibilities.

How is it (*pothen*), then, that heat comes to be in the air, if the sun is not hot and rays are not bodies? For if they are not / corporeal, there is no friction. But if there is no friction, how will the air be warmed? For that friction can, when it is very intense (*an puknotera genêtai*), warm to such an extent as to ignite [something] is evident from manifest facts. And even from wood, and from rocks and iron, when they are rubbed against one another, kindling occurs. And in the case of burning mirrors, if we did not position the ray [of light] in relation to the fuel in such a way that // the reflection of rays [at the mirror] makes not an obtuse but an acute angle, so that the doubling [of the ray] ends nearly in a straight line, and by its closeness, rubs together the air trapped in between,.kindling would not occur, and this we readily perceive. So that the appearances lend support to those who suppose (*hupothemenois*) the rays are bodies / and not to those who suppose they are actualities. For how does the actuality of what is not hot warm? And how could friction come about by the agency of the incorporeal?

In answer to this, I say that just as, due to soul, which is not hot, a certain life-giving actuality is produced in a body which arouses the inborn heat and engenders the living creature, and whenever the soul is absent, straightway the inborn heat / is extinguished--in this way also I say that, from the sun, a certain life-giving actuality is produced by means of (*dia*) the light in the air, and this [actuality] moves the inborn heat of the air and warms it. And just as the passionate (*thermoeidês*) power of the soul, though not itself warm, whenever it is moved, heats the blood around the heart, and similarly also thought, though it is an incorporeal actuality of soul, warms, / in this way it is not unreasonable that the sun also, though not warm, by the life-giving actuality of it, which is light, moves the quality of warmth in the air and heats it. And if the actuality is capable of warming, where there is more actuality, the warming will necessarily be greater, so that when the actuality is doubled in reflections and little air is enclosed inside, seeing that little air is warmed by more actuality /--not by friction but in the way described, by moving its inborn quality of heat--it is reasonable that kindling occurs.

We shall again give the same reasons for midday being warmed more than dawn and dusk. Because of the fact that, when the sun rises and sets, the actualities, in being reflected, make obtuse angles and / trap little air inside, for this reason they warm less. Moreover, at both dawn and dusk, the sun strikes us through more mist; and the mist is watery (for it is vapor), and for this reason, it is reasonable that the actualities striking us through more mist warm less. And that the mist is greater at sunrise and sunset than at noon is evident / from the following: For let the inside circle be the earth, and let the outside circle be a circle enclosing the earth in mist. It is evident, then, that the lines extending from the center of the earth will be equal to one another, since for every circle, the straight lines extending to the circumference are equal to one another. But if you extend straight lines from our portion of the earth and [if you extend them] to the heavens, it is evident that, because the / whole earth has the role of [being] the center in relation to the universe, even if these lines are unequal, nevertheless, because the difference among them is imperceptible considered in relation to the magnitude of the universe, they will be equal. But if you extend the straight lines

from our region [of the earth] to the mist, since as was shown in
333.1 the *Metereology*, the mist is not carried to a great height // above the earth (anyway, there are mountains above the clouds) nor do we live at the center [of the earth], of necessity the lines will not be equal to each other, but the ones at zenith will be shortest, and of the rest, those farthest from zenith will be longest and those nearer to zenith, shorter. / And of the straight lines drawn from the surface [of the earth] where we live, the one at zenith is shortest and the one through the center is longest, and of the rest, those nearer zenith are shorter and the remainder [are] longer. And in summer,[6] when [the sun] spends even more time above the earth, it warms more. For light is naturally inclined, whenever it falls on some smooth and glittering body / to act on it--e.g. on silver and mirrors and water and glass and many other bodies--in such a way that it also itself sends back the similar actuality (*tên homoian energeian antipempein*). Every body, then, that receives the actuality of light is also naturally inclined itself to send back the same [actuality], but especially smooth bodies, like those mentioned. For this reason, then, the moon also receives the light of the sun and illuminates the things / here below, and we call this reflection (*antanaklasin*) of light, the [reaction] from things that have received [light] and are able to act with a similar action (*homoian energein energeian*). In summer, then, and at midday, when there are more similar actualities (*pleionôn energeiôn homoiôn*) in the same place, the air happens to be warmed more. In this way, then, do they resolve the difficulties brought against the account concerning actualities.[7]

And if light is an actuality and not / a body, and further if vision does not take place by the emission of visual rays which are bodies but the actualities of visible objects come to us, why in the world did Aristotle use these hypotheses in the *Meteorology*? I say, then, he used the hypotheses of [light] rays and visual rays because they are clearer. For it is not easy to conceive (*ennoêsai*) actualities being reflected or, in general, [to conceive] the actualities of colors passing / through the air. But as many things as are established (*kateskeuasthê*) by the hypotheses [of visual rays], the same things also shall be established by means of these. For as we said there that cones go out from the eyes from a certain center, then widen as they go forward, evidently it will be said thus also in the case of actualities. For since the actualities of visible objects are discriminated (*krinontai*) in a certain narrow center of the crystalline, / they must originate from the width of the visible object itself and terminate in the narrow eye. And this is a cone having as apex the pupil or the center of the crystalline, and as base, the visible object itself. So that, with respect to the things demonstrated, the difference between the hypotheses does no harm. And [what is demonstrated] concerning reflections will hold similarly for both hypotheses. / These [arguments], then, are sufficient to argue for the Aristotelian hypotheses.

But the account in terms of (*peri*) actualities seems difficult to imagine--how it is possible that actualities, something incorporeal (*asômaton ti pragma*), are both reflected and make angles. For obtuse and acute angles have to do with separation, and separation // pertains to magnitude, and magnitude pertains to bodies and is in bodies. But actuality is incorporeal, having existence not in the air but in what is being actualized. How, then, is it reflected apart from its own substratum? Moreover, if it is reflected, either it is reflected along with the air or by itself. If, then, it is reflected by itself, again how / will something incorporeal be reflected and make angles? But if [it is reflected] along with the air, since the air, being moist, mutually replaces (*antiperiistatai*) itself, of necessity what is reflected will push another air whose place it occupies, and a different portion of air will occupy its place. It is evident, then, that the air displaced by the reflected air, along with the actuality of light in it, will yield / to the reflected [air], so that the actuality of light in one and the same portion of air will not be doubled. The other actuality yields, with the replaced air, to the actuality reflected with the air.[8] But if the actuality is not doubled, it does not warm more. Evidently, the explanation of the reflections of actualities is useless [to account] for production of greater warmth, whenever the actuality that is first in the place where the reflection occurred is replaced by the actuality / reflected with the air.

But against this, I say just what I have already said, that in the fashion by which the moon, when it has received the actualities of the sun and is illuminated because of the quality of brilliance of its body, is also naturally inclined itself to act in return (*antenergein*) and illuminate things here below, in this way also most other [things act],[9] those that receive the / actualities of light and are naturally inclined to actualize them in return, like water, glass, silver, bronze, and simply all smooth and glittering bodies. Whenever it is said, then, that actuality is reflected from smooth bodies, nothing else is said than that such bodies, when they receive the actualities of both light and visible objects, do not reflect (*poiei klasthai*) the numerically same actualities which they received, but that / they are themselves naturally inclined to produce such actualities in return (*antenergein toiautas energeias*). And for this reason, we pay close attention when using burning mirrors, so both the actuality from the sun falling on the mirror and the [actuality] generated in return (*palin ginomenên*) are not far apart but close together, and due to this closeness, produce warmth. The same [accounts] apply also to visual rays. /

Let it be decided, then, in this way. The former difficulty one might reasonably raise: if seeing does not take place by the emission of visual rays but the actualities of visible objects come to the eye, how do we know the distances between us and the things seen? For if the actualities come to the eye, both things near and things far ought to be seen in like manner. / The fact that we know the distances of visible objects shows that their images (*eidôla*) or actualities do not come to us. But how could one even imagine actualities of shapes and colors going through the air? But the Philosopher said in answer to this that the actualities do not go up to (*phoitan mechri*) the eye, but simply the whole air is filled full with all visible objects. / But this is still more puzzling and more absurd. For in the first place, the same difficulty remains. For if the air is filled in common with the
335.1 images (*emphaseôn*) of visible objects, // how do we know distances? For both the exceedingly distant and the near ought to appear in like manner. And even the most distant things ought not to elude perception, if it is true that all the air is filled with all images. But perhaps he would answer this by saying that, just as those who hypothesize visual rays / say that [the rays] weaken as they proceed to a distance and, for this reason, we do not see things far away, the same thing also we suppose in the case of actualities: [that they] weaken as they proceed very far. But against the former difficulty, what would he say? For if the whole

air is filled with images, what necessity is there for supposing that actualities go along straight lines? For we ought to see the things themselves that are not far from us and simply / whatever we are capable of seeing wherever we might look, if the whole air is filled with images. Why, then, do we not see the heavens and every visible object without looking at them?

And it is possible, arguing on behalf of the Aristotelian doctrine (*doxês*), to call to witness manifest facts themselves. For we see that, whenever a ray of sunlight goes through glass colored by red or some other such / color, the air, while remaining unaffected, transmits (*diaporthmeuei*) both the color and shape of the glass until it touches (*propsausêi*) some solid body and, finally, the color and shape of the glass are impressed (*enapomattetai*) on it. And, in truth, it is marvelous to see. For someone who is looking into the air through which the ray of sunlight passes sees the air unaffected by / the color and shape of the glass. Nevertheless, on the solid body where the ray terminates, both the color and the shape are seen. And it is not even possible, I presume,[10] to attribute [this result] to the reflection of visible rays there. For such a thing happens not simply on smooth bodies but also on wool, if the ray falls there, and on cloth and sand and / anything whatsoever, and even when the eye happens to be in a position so that there could be no reflection from there to the glass. It is clear, then, henceforth that, in truth, the actualities of visible objects, when they pass through the air without effect and thus are produced (*ginomenai*) in the organ of perception, because [they produce such an effect] in a solid body, there impress (*entupousi*) both the colors and shapes of visible objects. And when the organ of perception is acted upon by them in such a way, / the discrimination of the visible objects takes place in perception (*hê krisis epi tên aisthêsin diabainei*).

But it is possible to say against this that we do not see the same thing take place in the case of visible objects that happens in the case of the example. For what was said--[color appearing on the solid body]--does not occur unless the ray falls through the glass, while visible objects are seen even when no ray falls on them along a straight line from the sun. Moreover, in the case of the / [colored] glass, an image (*phantasia*) of the color appears on those things on which the ray falls, but no such image appears in

the case of visible objects. For it appears that [visible objects] do
336.1 not act at all // on other bodies, even if the ray passes through them (*di'autôn ... pherêtai*). What takes place in the case of the example is clearly different from the actuality of visible objects or simple perception (*haplôs lêpsin*).

But this difficulty also is raised against the account given in terms of actualities. Why in the world, they say, whenever a certain stain (*chumos*) is floating in the eye, do we not see that it / is inside? For if, in general, the actualities of perceptible objects act (*drôsei*) when they have been produced (*genomenai*) in the eye itself, and our visual rays do not arrive at them, we ought to see the stain constituted in the eye is inside. But, as it is, this does not happen, but we perceive it [to be] outside. Indeed, if that stain happens to be colored, we perceive the color outside in the air. / And if it has a shape, we in turn perceive the shape also outside. And no one would say that after its actuality has gone outside into the air, it is reflected back into the eye. For it is not the nature of visual rays or actualities to be reflected just by the air, and that too since the air is sometimes very fine. Moreover, if this were the case, why would not the color of the iris / make [the air] the same color. For the actuality of [the iris], when it has gone out and been reflected toward us, ought to be perceptible (*antilêptên*) to us. But, as it is, it is not. It is well-evident, then, they say, that the optical *pneuma*, when it has been acted upon by the stain inside, then goes outside and, making use of the illuminated air as a vehicle (*ochêmati*), thus perceives the affection produced in it by the stain inside. / For since the *pneuma* is continuous in itself, when that part striking the stain is once acted upon and altered, it acquires (*ischei*) the alteration successively by continuity. And since the perception of visible objects takes place in no other way than through the illuminated air, it is reasonable that it perceives the affection as occurring outside.

Against these things, then, we repeat our former position: / that we will not first accept that visual rays go out from the eyes until they resolve the impossibilities that follow from that hypothesis. For if [the hypothesis] of rays going out of the eye is completely confuted, we shall not, I suppose, suffer them to give arguments for the questions before us while using this hypothesis as true. We say, then, following [our] hypotheses, mentioned

before, / that this stain constituted in the eye is found only around the aperture itself of the iris outside the humors. For if it were constituted farther inside, either in the humors or in the optical nerve, it would altogether impede vision. We say, then, in truth, that the optical *pneuma* goes down from the brain through the optical nerves. It comes out to the crystalline, / and in the [crystalline] are its terminations. For which reason also, the discrimination (*hai kriseis*) of visible objects takes place there. For which reason too, it is transparent, in order that the actualities of visible objects might be transmitted through it to the optical
337.1 *pneuma*. For which very reason // too, the area inside the iris is pierced. For since the iris is colored and, for this reason, is likely to impede transmission of the actualities of colors, the part [of the iris] in front of the crystalline is pierced, in order that the actualities of visible objects may pass unimpeded through the cornea and the humors, which are transparent / and colorless. Since, then, the stain is constituted outside the humors and since the humors are transparent, the actuality of the constituted color or shape comes through them to the optical nerve. And since, even if the fluids are transparent, nevertheless it is by the presence of light that their transparency is perfected and becomes actually transparent, being potentially [transparent] before, for this reason the fact that we do not see when we close our eyes / cannot tell against our position.

For what reason, then, do we not see also the color of the iris, when its actuality shows through the humors? For we already said that the part [of the iris] in front of the crystalline is pierced. Since [the actuality] of it, then, is not in that part of the substance in which visible objects are discriminated, because the terminations of the optical *pneuma* are there, / it is reasonable that perception of it does not occur. "What then?" he says, "does the optical pneuma occupy a place only as large as the size of the pupil? Then, even if it occupies so small a place and remains unmoved in that part, is it not reasonable that it is diffused on either side? And does not the growth and diminution of this make the whole eye either more concave or more projecting?" My brief answer / to these objections is that, even if diffusion of the *pneuma* on either side happens to occur, nevertheless perception of visible objects occurs in only that part where the pupil is situated. And this is clear both from manifest facts themselves and from reason. For we see that, if some affection is produced in the pupil itself--I mean, either the formation of a stain or some foulness (*hrupos tis*) and / coarsening (*pachunsis*) of the cornea--straightway, our vision is impeded. Nevertheless, if the [pupil] itself and the parts around it are healthy (*kata physin echoi*), while something affects all the parts on either side of it--either a tumor (*skirron*) or a stain floating or some such thing--vision occurs unimpeded, since evidently perception occurs through the pupil alone. And the reason is evident. For if it is for this reason that the iris is pierced there--in order that its own color will not / impede the reception of visible objects--wherever it is not so affected, it evidently will impede vision. So, even if the *pneuma* is diffused through the whole eye, it is impossible that it perceive through the whole eye. And just as, if a certain house has only one window, and the glass in that window is deeply black, since light cannot illuminate the house through the window, those inside perceive neither the window or anything else, in just this way also does it happen in the iris. For even if everything inside of the eye is transparent, // since light cannot, due to the color of the iris, lead the transparency in them into actuality, the power of those [parts] inside will not perceive even the iris itself. And to put it quite generally, if we do not see through the whole iris but through the pupil alone, it is useless to seek for what reason vision does not perceive also the iris. /

"Yes," he says, "And if we see the stain inside, without visible rays proceeding outside, why do we not perceive that the affection is inside, but we think the affection inside is outside?" I say, then, in answer to this, that nearly all conditions contrary to nature are naturally subject to such an [interpretation]. Indeed, those suffering from dizziness attribute their affliction to the things outside, / and this is not, I suppose, because a certain *pneuma* goes out from the brain to the outside. But even those having some flavor contrary to nature in the mouth think this belongs to the food and drink. Whence it was said and again shall be said that for something to be capable of perceiving certain things, it must not have been affected by any of the things it is to perceive; and similarly, what is capable of transmitting these / must not possess beforehand any of these qualities. And for this reason, since the air is capable of transmitting colors and sounds and smells, it is in itself (*kath'auto*) colorless and without sound and odorless. For if it possessed beforehand some color, this would impede its transmission of the rest. And for this reason, as I already said, colored glass is not able to transmit / all colors. Just as, then, glass colored by red or blue dye shows everything seen through it as similar in hue to itself, in this same way also whenever one of the faculties of perception is affected beforehand by one of the qualities it is capable of discriminating (*kritikon*), it imparts (*peritithêsi*) this quality to all other perceptible objects. In this way, then, / also the tongue, whenever it is acted upon, thinks this affection which it possessed beforehand belongs to all it tastes. For when it tastes something, since it possesses a stronger flavor in itself, it perceives that flavor first and thinks it accompanies what entered [the mouth]. Hence, people with this condition take stronger flavors, e.g. vinegar or some such thing, in order that / this [flavor], being stronger than the one residing contrary to nature in the tongue, may be perceptible to the potential faculty of perception.

The same thing, then, occurs also in the case of vision. For due to there being a color contrary to nature in us, we think the things outside are colored--as in the case of those afflicted by jaundice, for they think everything is yellow. <For> since the transparent parts of the eye, through which we perceive visible objects, / like the humors and the cornea, have become yellow, we think all the things we see through them are yellow, just as also the things seen through colored glass seem to have the same color as the glass. And in this way too, in the stain formed in the eye, we attribute its color and shape to the air. And just as I already said--just as in the case of the other organs of perception, // whenever they are in a condition contrary to nature, their affection seems to belong to perceptible objects, although perceiving does not take place in their cases by an emission--if it is reasonable also in the case of vision that the same thing happens, clearly we did not accept that, because the organ of perception perceives its own affection outside, *pneumata* go out of the eye. / But also in relation to the shape of the afflicting stain, it happens also to be seen outside for this reason: for if the stain, when it has taken effect, takes over (*apolabêi*) the whole pupil, all activity (*energeia*) is impaired. But whenever it blocks the optical power in a certain part of the pupil, since visible objects and light enter from every direction, but there is loss of sight in that part / due to the blockage by the stain, an illusion occurs (*planê tis ginetai*), and the part in which there is no vision seems to be a body outside, visible and similarly shaped. For which reason, even if the stain happens not to remain stationary but moves, it seems that outside too some body is always moving, when the loss of sight occurs in a different part. But why is this surprising, if there are mirrors which produce the appearance of images outside (*exô tên emphasin tôn eidôlôn parechonta*)? And yet, / would anyone think (*huponoêsoi*) the image appearing outside moves around in this same way as [we do think] occurs in the case of conditions contrary to nature in the eye?

For what reason, then, they say do we no longer see when excessive dilation of the pupil (*platukoriaseôs*) occurs? For on the one hand, those who believe in visual rays have a reasonable explanation: that the rays, on going out, are straightway scattered and diffused and do not preserve in emission / the shape which they have when the pupil is in a natural condition. But what would you say in the case of actualities? For when the pupil is wider, one ought to see better, since the actualities enter more easily. Again, then, it is possible to reply from the same hypotheses that, as has often been said, the perceptible objects must lie on a straight line with the organ of perception. Someone will say, then, that the actualities are diffused (*cheisthai*) / when the pupil is flat and are not carried to the center itself, where discrimination (*krisis*) takes place, but this is ridiculous. For if they were bodies, what is said might be plausible, but as it is, how could immaterial actualities be diffused? For no doubt, in whatever position the centers may be, there will be some visible objects on a straight line with the organ of perception in which discrimination takes place. What then / prevents the perception of them from being accurate? But what also is the reason--in the pupils being displaced and not being on one and the same straight line, either due to a natural error or by design, as whenever sticking a finger into the lower eyelid, we turn the eyeball upward while maintaining the position of the other eye--[what is the reason] due to such dispositions that we see things double? /

I, then, was once asked the cause of this and, making use of the hypothesis of visual rays, gave an account of the occurrence in this way: [Let us] say two cones go out from the eyes from one and the same undeflected straight line, which goes between the centers of the crystalline humors, and visual rays contact visible objects and make a triangle, having the perceptible object as a
340.1 base // and as sides the visual rays equidistant from the axis of the cone on each side, by which visual rays the discrimination of visible objects occurs, and there are two axes, one to each cone, around which are the remaining rays. If [we say these things], it is evident, I suppose, that if the straight line drawn through the centers of the crystallines is undeflected, / the visual rays from the two cones that are equidistant from the axes contact visible objects at the same points, so that the ends of the visual rays fit onto (*epharmozein*) one another and make equal angles. Since perceptible objects are judged by the angles made by the visual rays equidistant from the axes on each side, and since the cones are undeflected, / the rays equidistant from the axis of each cone contact the same points, so it is reasonable that they perceive a single thing as single. Just as, then, if there are two perceptible objects, due to the fact that different visual rays fall on them and they touch one another at no point and make different angles, for this reason also visible objects appear as two or more, similarly, if / the position of the rays is turned aside, it happens that from each cone, visual rays that are not equidistant from their axes contact the same visible object. And since the angles made from the same points of the visible object are different, it is reasonable that the judgements (*kriseis*) also are two, and for this reason, the one perceptible object seems to be two. In this way, then, from the hypothesis of visible rays, it is possible plausibly to give an explanation of what happens. / But what would someone say is the cause of this, using the hypothesis of actualities?

And in answer to the difficulty concerning distances--I mean the one that says, "If actualities come to us and nothing is sent out from the eyes to perceptible objects, how is it that we have perception of distances and magnitudes? For nothing would seem to be nearer or farther than any other thing, / when all perceptible objects come in the same way to the organ of perception. And how could any appear greater or smaller if they all terminate in one point?"--this [difficulty], then, they solve and say that if we perceive clearly in the case of hearing, whether what produces sound is distant or not, although clearly the perceptible object comes to the organ of perception and nothing / goes out from the faculty of perception, why is it surprising if this happens also in the case of vision, when the actuality of visible objects falls on the eyes? That the air which is struck and produces sounds does not itself draw near to hearing but by its power of conducting (*diêchê*) sounds, the air transports (*diabibazei*) sound as also the actualities of colors, is evident from what we perceive when we are in water / and sound is produced in the air--surely not the sound-producing (*psophountos*) air passing through the water, but the sound-conducting power of the water (for the ability to conduct sound is in water too), which is capable of reporting

341.1 (*apaggeltikês*) sounds. And yet, // even if the sound-producing air itself drew near the organ of perception, it would support the foregoing proposition even more. For how is it that, when the sound-producing object comes up to the organ of perception, we perceive whether the sound-producer is near or far?[11] So that, even if one cannot give the reason for such as this, nevertheless, that the hypothesis of Aristotle concerning visible objects is not / refuted by it, the matter of sound shows sufficiently at least. And in these cases, the power of the air to conduct sounds reports sounds to the organ of perception, when nothing goes out of the organ of perception, and equally--whether the sound-producing object is far or near, in front or behind--we perceive. /

"And in those in which this [the transparent] is potentially, there is also darkness." For since the transparent is not always actually transparent (i.e., the transparent in the sublunary world, for that of the heavens is always transparent), in those in which the potentially transparent exists, in these of necessity [there is] also darkness. For before it becomes actually transparent, there is darkness. For what is still potential according to such a / potentiality--i.e., the first--is in privation of that of which it is capable, and the privation of light is darkness. The absence, then, of those things whose presence produces light in the potentially transparent produces privation of light in this same thing, and privation of light is darkness. Just as vision, then, is a certain form, and the absence of this does not bring on (*epipherei*) some other form but only the privation of vision, / and this is blindness, and just as the absence of the helmsman sinks the ship, when no other form supervenes in the ship but there is only privation of the helmsman, in this same way also, the absence of light, of necessity, is darkness. And if light is the opposite of darkness, and darkness is privation, light of necessity is form. For the opposite of privation is form, / according to what was said in the *Physics*. But if light is form, and forms are incorporeal (for the [composite] of form and matter is body), it is clearly evident that light is not body.

But if someone did not wish light to be the opposite of darkness as state and privation but as contraries, for the reason that state changes into privation but privation does not change into state, but light and darkness / change into one another, and he wishes them to be contraries for this reason, since contraries change into one another, in this way too light will be shown to be no less incorporeal. For if light is the contrary of darkness, and in composite substances, no substance is contrary to any other as was shown in the *Categories*, and bodies are composite substance, clearly light is not a body. / For nothing is contrary to body, since nothing is contrary to a composite. For if fire is the opposite of water, it is [its opposite] not *qua* body but *qua* qualified by contrary qualities. So that darkness in one way is contrary, but in another way it is not contrary. Similarly, in one way it is privation, but in another way it is not. For if someone says that privation is what does not convert back into the form, // darkness is not privation but contrary. But if one were to say that privation, as was said in the *Physics*, is only the absence of form, which also converts back into form, darkness would be privation. Again, if someone were to say that those things qualified by form and possessing their own potentiality and actuality are contraries, darkness would not be the contrary of light, / for it has not form nor its own potentiality or actuality. But if someone were to say that simply all those things that change into one another are contraries, darkness would be the contrary of light. But that darkness is not form, is evident also from what follows: the cause of light is the sun or fire or the moon or some other such thing, but we see no cause of darkness, but the absence of a light source alone suffices / for there to be darkness. But darkness is not even a certain quality of the air, as one might suppose. For if it is a quality of air, why is it not seen in light? And how does it disappear all at once and without the passage of time, due to the presence of light? For no quality departs from its substratum without the passage of time. But if darkness is the privation of the actuality of a light source, / light clearly is actuality. But no actuality is body. So clearly, light is not a body.

Notes

1. The text may be corrupt (line 3), but it is clear from what follows that Philoponus was arguing against the possibility that light or visual rays are corporeal.

2. For the corrupt 10*tôn*-11*einai*, supplying *exagônou de pleura* before *tôn* and reading *esti* for *einai*. See Hayduck's apparatus and Sophonias, *In de Anima Paraphrasis* (vol.23, part 1 of *Comm. in Arist. Gr.*, ed. by M. Hayduck), 76.30.

3. A mistake in the Greek; it takes four hours to cover two signs of the zodiac.

4. *On Generation and Corruption* I.8.

5. Applied to cases of true actuality, the meaning of this cognate accusative is both that the same sort of 'change' occurs in each part of the medium and that the new state of the part actualized is the same as the state of what has actualized it. Neither possible translation, "to act with a similar action" or "to actualize the similar actuality," captures all of this meaning.

6. In line 8, reading *therei* against the edition's *thermôi*; compare line 16. Either *thermôi* means 'the hot season,' in this context, or it is a corruption of *therei.*

7. I have translated *epiluontai*, the editor's choice over the alternate readings given in the apparatus. See his indication of problems in the ordering of the text. Philoponus does not say who the other thinkers are who hold to "the account concerning actualities." Compare 340.27.

8. In line 11, I read *meta* instead of *metaxu.* Compare line 5.

9. In order to render lines 17-21 a complete sentence, I have not translated *de* in line 19.

10. In line 22, accepting the reading in the Trincavellus edition (Venice, 1535) and reading *oude pou*, against Hayduck's *ou dei pou.*

11. In line 3, accepting the suggestion of Hayduck and inserting *eite eggus* before *eite porrô*.

Appendix II

On the Interpretation of Physics *VIII.3, 253b15-35*

My understanding of Aristotle's analysis of motion owes much to Owen's treatment of *Physics* VIII.8, 263b9-26. However, my conclusions in Chapter Three led me to a different interpretation of *Physics* VIII.3, 253b15-35 from Owen's.[1]

Aristotle always insisted that there is no time of 'becoming' separate from being in one extreme or the other of change. Owen said the motivation for this viewpoint arises from Plato's puzzle that if change is from not-A to A, it must take place in an instant.[2] Plato relied on the law of excluded middle to maintain that there is no time in which a thing can be neither A nor not-A; either the thing has already changed or it has not. In Owen's account, Aristotle took the puzzle even further, saying that motion cannot even be in an instant, since there is no moment in which a thing can be neither A nor not-A. Aristotle, unlike Plato, went on to solve the problem (263b9-264a7) in a way which Owen summarized as follows:

> In brief, Aristotle takes the puzzle to show that it is a mistake to look for a special time-reference such that the subject is neither white or not-white. The primary moment at which the subject becomes (or, as Aristotle prefers to say, has become) white is the first moment at which it is white. And given this moment, it becomes improper to talk of the last moment at which the subject was *not* white, for the two moments would have to be consecutive. Equally, given a last moment of stability there cannot be a first moment of change.[3]

This brief summary provides an important standard for comparison, because Owen seems to think that some formulations of this solution in the *Physics* are less adequate than others.[4] Indeed, he thought that the treatment of *athroa alloiôsis* in 253b21-30 is one example of Aristotle's preoccupation with Plato's 'not-A to A' problem and he believed this is a passage where Aristotle "agrees that some changes take no time at all."[5] However, a comparison of the examples involved with their use in other parts of Aristotle's work will show that he did not intend to relieve *athroa alloiôsis* of its expected correlation to a time period.

Owen based his interpretation of this passage on the example of regaining health (253b26-28) that follows the example of freezing:

> Among other instances he cites the recovery of health, which is "a change to health and to nothing else"; in other words, although the process toward recovery may take time the actual recovery is simply the change from not-A to A.[6]

According to this interpretation, we obtain three terms in a process of change:

hugiazein	*hugiainein*	*hugieia*
becoming healthy	recovering health	health
not-A	change that takes no time	A

The specification of the middle terms is easier for this triad than it would be for some others, like freezing, due to the existence of two separate words in Greek for 'becoming healthy' and 'recovering health.' Owen saw *Metaphysics* Θ.6, 1048b25, where these two words are coupled, as an example of a movement culminating in a change which takes no time.[7]

In *Metaphysics* Θ.6, *hugiazein* and *hugiainein* are coupled to parallel the coupling of the present and perfect tenses of other verbs, like *manthanei* and *memathêken*. What *memathêken* and *hugiastai*, 'has learned' and 'recover health,' have in common is the achievement of the goal implied by the action. One cannot use the present and perfect tenses at the same time to describe a motion. Only when the motion has ended may the perfect tense describe the activity. Thus, *memathêken* and *hugiastai* refer to motion that is complete, an end which has been attained. Consequently, the end, *hugieia*, has already been attained when *hugiastai* may be used. This point is made in more precise terms in *Physics* VI.5, 235b19-236a7. First, Aristotle argued (b19-20) that what has changed, when it has changed, is in that toward which the motion was tending. If it were not, it would be somewhere on the way to the goal and so, when it *had changed*, would still *be changing*, which is impossible. He then showed that the first 'when' in which something has changed is indivisible (b30-236a7). Furthermore, while there is a first moment of 'has changed,' there is no first moment of 'changing.' By the first moment of 'has changed,' he meant the first moment in which the change is complete (236a8). This passage confirms the point made in *Metaphysics* Θ.6 about the meaning of the perfect tense in describing motion, and the same point applies as well to *hugiastai*. Owen himself stated the relation between the perfect tense and the completion of motion, in presenting Aristotle's solution to Plato's problem about change: "The primary moment at which the subject becomes (or, as Aristotle prefers to say, has become) white is the first moment at which it is white."[8] There is no reason to think that this analysis would not also apply to Aristotle's consideration of *athroa alloiôsis* in 253b23-30.

The crucial sentence in Owen's interpretation is : "the change must be to health and to nothing else"(a28). He saw this sentence as expressing a change from not-A to A. However, the reference to regaining health forms a unit with the two sentences which follow it, the whole comprising a third argument against continuous change (a26-33). The first argument states the necessity for an interval between growth and diminution in the same subject (a13-14). The second and longest argument states that change, whether in place, quantity, or quality, may not always be continuous with respect to time (a14-25). The third argument explains that alteration is always change to an opposite and takes a certain time (a26-33). But being change to one particular thing and not something else, it cannot go on forever, just as a rock does not become harder or softer without limit, but simply becomes hard. A fourth argument against continual motion follows, based on the idea that some things must be in their natural places (a33-36). It seems clear that the example of health should be separated from the treatment of *athroa alloiôsis* that precedes it, due to its introduction by the particle *eti* signifying the beginning of an additional argument. The fourth argument which follows the example of health is also introduced by *eti*.

Owen's example of a movement leading up to a change that takes no time harmonizes nicely with Aristotle's general analysis of motion. The presence of this example in Aristotle's consideration of *athroa alloiôsis* forms part of an argument against the indefinite continuability of motion, due to the fact that alteration must terminate in its opposite. In the final analysis, there is no justification, based on 253b21-30, for thinking that alteration takes place in no time whatsoever.

Notes

1. Owen, "*Tithenai ta phainomena,*" 185-86.

2. Plato, *Parmenides* 156c1-8.

3. "*Tithenai,*" 186-67.

4. "*Tithenai,*" 186, n.96.

5. "*Tithenai,*" 185.

6. "*Tithenai,*" 185-86.

7. "*Tithenai,*" 186, n.93.

8. "*Tithenai,*" 186.

Selected Bibliography

I. Primary sources

A. Ancient texts

Aristotle. *De Anima*. Ed. W.D. Ross. Oxford: Clarendon Press, 1956.

______. *Metaphysica*. Ed. W. Jaeger. Oxford: Clarendon Press, 1957.

______. *Nicomachean Ethics*. Ed. I. Bywater. Oxford: Clarendon Press, 1890.

______. *On Sense and Sensible Objects. Parva Naturalia*. Loeb Classical Library. Cambridge: Harvard University Press, 1934.

______. *Physica*. Ed. W.D. Ross. Oxford: Clarendon Press, 1956.

______. *Prior and Posterior Analytics*. Edited with introduction and commentary by W.D.Ross. Oxford: Clarendon Pres, 1949.

Damascius. *Reliquiae Vitae Isidori*. Edited and annotated by C. Zintzen. Hildesheim: Georg Olms Verlagsbuchhandlung, 1967.

Euclid. *Optics. Opera Omnia* VII. Ed. by I.L. Heiberg and H. Menge. Leipzig: B.G. Teubner, 1895.

Migne, J.-P., ed. *Patrologiae Cursus Completus, Series Graeca* LXXXV. Paris: L. Migne, 1860.

Plato. *Timaeus. Opera* IV. Ed. by J. Burnet. Oxford: Clarendon Press, 1902.

Ptolemy. *L'Optique de Claude Ptolémée dans la version latine d'après l'arabe de l'émir Eugène de Sicile*. Edited with commentary by Albert Lejeune. Louvain: Bibliothèque de l'Université, Bureaux de recueil, 1956.

Theon of Alexandria. *Recensio Opticorum. Opera Omnia* VII. Ed. I.L. Heiberg. Leipzig: B.G. Teubner, 1895.

B. Ancient commentaries

Commentaria in Aristotelem Graeca. Edited under the authority of the Prussian Academy of Letters. Berlin: G.Reimer, 1882-1909.

I. Alexander of Aphrodisias. *In Aristotelis Metaphysica*. Ed. M. Hayduck, 1891.

III.1 ______________. *In librum De Sensu*. Ed. P. Wendland, 1901.

VIII. Simplicius. *In Aristotelis Categorias*. Ed. C. Kalbfleisch, 1907.

IX. ______. *In Aristotelis Physicorum libros quattuor priores*. Ed. H. Diels, 1882.

X. __________. *In Aristotelis Physicorum libros quattuor posteriores*. Ed. H. Diels, 1895.

XIII.1 Joannes Philoponus (Ammonius). *In Aristotelis Categorias*. Ed. A. Busse, 1898.

XIII.3 ____________. *In Aristotelis Analytica Posteriora cum Anonymo in librum II*. Ed. M. Wallies, 1909.

XV. ______________. *In Aristotelis De Anima libros*. Ed. M. Hayduck, 1897.

XVI. _______________. *In Aristotelis Physicorum libros tres priores*. Ed. H. Vitelli, 1887.

XVII. _____________. *In Aristotelis Physicorum libros quinque posteriores*. Ed. H. Vitelli, 1888.

Supplementum Aristotelicum. Edited by the Prussian Academy of Letters. Berlin: G. Reimer, 1885-1903.

II.1 Alexander of Aphrodisias. *De Anima liber cum Mantissa*. Ed. I. Bruns, 1887.

Olympiodorus. *Commentary on the First Alcibiades of Plato*. Ed. L.G. Westerink. Amsterdam: North-Holland Publishing Col, 1956.

Proclus. *In Parmenidem. Opera inedita*. Compiled and emended by V. Cousin. Paris: A. Durand, 1864.

_____. *In Platonis Timaeum commentaria*. Ed. E. Diehl. Leipzig: B.G. Teubner, 1903-6, I-III.

_____. *In primum Euclidis Elementorum librum commentarii*. Ed. G. Friedlein. Leipzig: B.G. Teubner, 1873.

Westerink, L.G., ed. *Anonymous Prolegomena to Platonic Philosophy*. Amsterdam: North-Holland Publishing Co., 1962.

C. Medieval and early modern texts

Alhazen. "Le «Discours de la lumière» d'Ibn al'Haytham (Alhazen)." Trans. R. Rashed. *Revue d'histoire des sciences et de leurs application* 21, no.3 (Juillet-Septembre, 1968): 197-224.

Bacon, Roger. *Opus maius*. Ed. with introduction by J.H. Bridges. Frankfurt: Minerva, 1964. I-III.

Descartes, René. *Le Monde. Ouevres de Descartes* XI. Ed. C. Adam and P. Tannery. Paris: J. Vrin, 1957.

Grant, Edward, ed. *A Source Book in Medieval Science*. Cambridge: Harvard University Press, 1974.

Pecham, John. *John Pecham and the Science of Optics: Perspectiva communis*. Edited with translation and notes by David Lindberg. Madison: University of Wisconsin Press, 1970.

II. Secondary sources

Ackrill, J.L. "Aristotle's definitions of *psuchê*," *Articles on Aristotle*, ed. Barnes, J., Schofield, M. and Sorabji, R. London: Duckworth, 1979, vol. 4 (Psychology and Aesthetics).

________. "Aristotle's Distinction Between *Energeia* and *Kinêsis*." *New Essays on Plato and Aristotle*. Ed. Renford Bambrough. London: Routledge and Kegan Paul, 1965, 121-141.

________. "Aristotle's Theory of Definition: Some Questions on *Posterior Analytics* II.8-10." *Aristotle on Science: The Posterior Analytics*. Ed. Enrico Berti. Padova: Editrice Antenore, 1981, 359-384.

Annas, Julia, trans. *Aristotle's Metaphysics M and N*, with introduction and notes. Oxford: Clarendon Press, 1976.

Baynes, Norman H. "Alexandria and Constantinople: A Study in Ecclesiastical Diplomacy." *Journal of Egyptian Archeology* 12 (1926), 145-56.

Barnes, Jonathan, trans. *Aristotle's Posterior Analytics*, with notes. Oxford: Clarendon Press, 1975.

Barnes, Jonathan, Schofield, Malcolm, and Sorabji, Richard, ed. *Articles on Aristotle*. London: Duckworth, 1975-79. 4 vol.

Blumenthal, H.J. "Neoplatonic Elements in the *De Anima* Commentaries." *Phronesis* 21, no.1 (1976): 64-87.

Bolton, Robert. "Essentialism and Semantic Theory in Aristotle: *Posterior Analytics* II.7-10. *Philosophical Review* 85 (1976): 514-44.

Cameron, Alan. "The Last Days of the Academy at Athens." *Proceedings of the Cambridge Philological Soceity* 195 (N.S. 15) (1969): 7-29.

Cameron, Averil. *Agathias*. Oxford: Clarendon Press, 1970.

Clagett, Marshall. *Science of Mechanics in the Middle Ages*. Madison: University of Wisconsin Press, 1959.

Cooper, John. "Arisotle on the Ontology of the Sense." unpublished.

Cornford, Francis Macdonald, trans. *Plato's Cosmology: The Timaeus*, with commentary. Indianapolis: Bobbs-Merrill, 1959.

Courcelle, Pierre. *Les Lettres grecques en Occident*. Paris: E de Boccard, 1943.

De Groot, Jean. "Philoponus on *De Anima* II.5, *Physics* III.3, and the Propagation of Light." *Phronesis* 28, no.2 (1983): 177-96.

Demoss, David, and Devereux, Daniel. "Essence, Existence, and Nominal Definition in Aristotle's *Posterior Analytics* II.8-10." *Phronesis* 33, no.2 (1988): 133-54.

Dodds, E.R. *The Greeks and the Irrational*. Berkeley: University of California Press, 1951.

Evrard, Étienne. "Les convictions religieuses de Jean Philopon et la date de son commentaire aux «Meteorologiques»."*Bulletin de l'Academie Royale de Belgique*, Classe des Lettres 39 (1953): 299-357.

__________. "Jean Philopon, son commentaire sur Nicomaque et ses rapports avec Ammonius." *Revue des études grecques* 78 (1953): 592-98.

Frend, W.H.C. *The Rise of the Monophysite Movement*. Cambridge: Cambridge University Press, 1972.

Gersh, Stephen. *Kinêsis Akinêton: a study of spiritual motion in the philosophy of Proclus*. Leiden: E.J. Brill, 1973.

Gomez-Lobo, Alphonso. "The So-called Question of Existence in Aristotle, *An. Post.* 2.1-2."*Review of Metaphysics* 33 (1980): 71-89.

Grant, Edward. "Aristotle, Philoponus, Avempace and Galileo's Pisan Dynamics." *Centaurus* 11 (1965): 79-95.

Grant, Edward. "The Principle of the Impenetrability of Bodies in the History of Concepts of Separate Space from the Middle Ages to the Seventeenth Century." *Isis* 69, no. 249: 551-571.

Gudeman, A. "Ioannes Philoponus." *Real-Encyclopädie der klassichen Altertumswissenschaft*. Ed. A. Pauly, G. Wissowa, and W. Kroll. New edition. vol 9: 1764-95.

Hardie, W.F.R. *Aristotle's Ethical Theory*. Oxford: Clarendon Press, 1968.

Ingekamp, H.G. "Zur Stoischen Lehre von Sehen." *Rheinische Museum für Philologie* N.F. 114, no.3 (1971): 240-46

Irwin, Terence. *Aristotle's First Principles*. Oxford: Clarendon Press, 1990.

Lejeune, Albert. *Euclide et Ptolémée*. Louvain: Bibliothêque de l'Université, 1948.

Lemerle, Paul. *Le premier humanisme Byzantin; notes et remarques sur enseignement et cultur à Byzance des origines au Xe siècle.* Paris: Presses universitaires de France, 1971.

Lindberg, David C. *Theories of Vision from Al-Kindi to Kepler.* Chicago: University of Chicago Press, 1976.

Lloyd, G.E.R. "Saving the Appearances." *Classical Quarterly* 28, no.1 (1978): 202-22.

Mansion, Suzanne. *Le jugement d'existence chez Aristote* (Paris: Declée, 1946).

Mason, Hugh. *Greek Terms for Roman Institutions: A Lexicon and Analysis. American Studies in Papyrology* 32 (1974).

Merlan, Philip. "Ammonius Hermiae, Zacharias Scholasticus and Boethuis." *Greek, Roman, and Byzantine Studies* 9, no. 2 (1968): 193-203.

Miller, Fred D., Jr. "Did Aristotle Have a Concept of Identity?" *Philosophical Review* 82 (October, 1973): 483-90.

Moraux, Paul. *Alexandre d'Aphrodise, exégète de la noétique d'Aristote.* Paris: E. Droz, 1942.

Morrow, Glen R., trans. *A Commentary on the First Book of Euclid's Elements,* with introduction and notes. Princeton: Princeton University Press, 1970.

Mueller, Ian. "Aristotle on Geometrical Objects." *Archiv für Geschichte der Philosophie* 52, no. 2 (1970): 156-71 (reprinted in *Articles on Aristotle,* vol.3 (Metaphysics).

_______. "Aristotle's doctrine of abstraction in the commentators." *Aristotle Transformed.* Ed. Richard Sorabji. Ithaca, N.Y.: Cornell University Press, 1990, 463-480.

Nussbaum, Martha. *Aristotle's De Motu Animalium.* Princeton: Princeton University Press, 1978.

Owen, G.E.L. *Logic, Science, and Dialectic: Collected Papers in Greek Philosophy,* ed. Martha Nussbaum. Ithaca, N.Y.: Cornell University Press, 1986.

Owen, G.E.L. "*Tithenia ta phainomena.*" *Aristotle.* Ed. by J.M.E. Moravcsik, Garden City, N.Y.: Anchor Books, 1967, 167-90 (reprinted in *Articles on Aristotle,* vol. 1 (Science), and *Logic, Science, and Dialectic: Collected Papers in Greek Philosophy*).

Owens, Joseph. *The Doctrine of Being in Aristotle's Metaphysics,* 3rd. ed. Toronto: Pontifical Institute of Medieval Studies, 1978.

Penner, Terry. "Verbs and the Identity of Actions--a Philosophical Exercise in the Interpretation of Aristotle." *Ryle*. Ed. O.P. Wood and G. Pitcher. Garden City, N.Y.: Anchor Books, 393-460.

Richard, Marcel."*Apo Phônês*." *Byzantion* 20 (1950): 191-222.

Ross, W.D., ed. *Aristotle's De Anima*, with introduction and commentary. Oxford: Clarendon Press, 1961.

_________. *Aristotle's Metaphysics*, with introduction and commentary. Oxford: Clarendon Press, 1961, I-II.

_________. *Aristotle's Physics*, with introduction and commentary. Oxford: Clarendon Press, 1936.

Ryle, Gilbert. *Concept of Mind*. Harmondsworth, Middlesex: Penguin Books, Ltd., 1963.

_________. *Dilemmas*. Cambridge: Cambridge University Press, 1960.

Sabra, A.I. *Theories of Light from Descartes to Newton*. London: Oldbourne Book Co., 1967.

Saffrey, H.D. "Le chrétien Jean Philopon et la survivance de l'école d'Alexandrie au VIe siècle." *Revue des études grecques 67 (1954): 396-410.*

Sambursky, Samuel. *The Physical World of Late Antiquity*. London: Routledge and Kegan Paul, 1962.

Simon, Gerard. *Le regard, l'être, et l'apparence dans l'optique de l'antiquité*. Paris: Seuil, 1988.

Sokolowski, Robert. "Matter, Elements, and Substance in Aristotle," *Journal of the History of Philosophy* 8, no.3 (July, 1970): 263-88.

Sorabji, Richard."Aristotle on the Instant of Change." *Proceedings of the Aristotelian Society*, supplementary vol. 50 (1976): 69-89 (reprinted in *Articles on Aristotle*, vol. 3 (Metaphysics).

_________, ed. *Aristotle Transformed*. Ithaca, N.Y.: Cornell University Press, 1990.

_________. *Matter, Space, and Motion*. Ithaca, N.Y.: Cornell University Press, 1988.

_________, ed. *Philoponus and the Rejection of Aristotelian Science*. Ithaca, N.Y.: Cornell University Press, 1987.

Tannery, Paul. "Sur la période finale de la philosophie grecque." *Mémoires scientifiques* 7, 211-41.

Todd, Robert B. *Alexander of Aphrodisias on Stoic Physics*. Leiden: E.J. Brill, 1976.

_________. "*Sunentasis* and the Stoic Theory of Perception." *Gräzer Beitrage* 2 (1974):251-61.

Trouillard, Jean. "«Agir par son être même»: la causalité selon Proclus." *Revue des sciences religieuses* 32 (1958): 347-57.

Tsouyopoulos, Nelly. "Die Entstehung Physikalischer Terminologie aus der Neoplatonischen Metaphysik." *Archiv für Begriffsgeschichte* 13, no.1 (1969): 7-33.

Vasiliev, A.A. *Justin the First*. Cambridge: Harvard University Press, 1950.

Vendler, Zeno. "Verbs and Time."*Philosophical Review* 66 (April, 1957): 143-60.

Ver Eecke, Paul, trans. *L'Optique et Catoptrique*, with introduction and notes. Paris: de Brouwer et Cie, 1938.

Westerink, L.G. "Deux commentaires sur Nicomaque: Asclépius et Jean Philopon." *Revue des études grecques* 77 (1964): 526-35.

________. "Philosophy and Medicine in Late Antiquity." *Janus* 51 (1964): 169-77.

Wildberg, Christian. *Philoponus Against Aristotle on the Eternity of the World*. Ithaca, N.Y.: Cornell University Press, 1987.

Wolff, Michael. *Fallgesetz und Massebegriff: Zwei wissenschaftshistorische Untersuchungen zur Kosmologie des Johannes Philoponus*. Berlin: Walter de Gruyter, 1971.

Index Locorum

(References in endnotes are indicated by the page in the text in which the note number appears.)

Index Verborum